Optimal Knowledge Management:
Wisdom Management Systems Concepts and Applications

Table of Contents

Section II: Underlying Concepts of Optimal Knowledge Management/Wisdom Management Systems

Section IV: Applications of Optimal Knowledge Management/Wisdom Management Systems

Preface

Introduction to Wisdom

A starting point for wisdom is a humble assertion: "I (We) don't know." This assertion can be the real beginning of wisdom. Wisdom can be defined very simply as "the ability to judge soundly." Because business transactions per se do not inspire much wisdom in decision makers, wisdom comes from connecting these transactions to each other and their change over time so that sound judgments can be made. Wisdom requires an intuitive ability, born of experience, to look beyond current situations in order to recognize exceptional factors and anticipate unusual opportunities and outcomes. Largely untapped today, wisdom is a vital organization resource, accumulated through experience, and applied to everyday learning at work. Basically, wisdom is a personal capacity acquired through creative thinking and experience. From this perspective, there is a tendency to replace past hierarchical and functional roles with learning relationships that focus on wisdom as the foundation of the new organization.

Prior Knowledge Management, Business Intelligence, and Smart Business Systems

Today, there are a number of information systems that tell the organization's decision makers where they have been and, to a degree, where they are going, but not much about "what needs to be done" to grow an organization over time in an optimal manner. For example, knowledge management systems give decision makers information and knowledge about an organization's operations while business intelligence systems analyze the results of operations, that is, give decision makers a good understanding of an organization's operations. More recently, smart business systems focus on optimization of an organization's operations beforehand. Although all of these systems represent improvements over time, there is need for a fundamental shift or a new paradigm in how information systems are effectively used by decision makers. More specifically, there is a need to employ optimal knowledge management/wisdom management systems or simply optimal KM/WM systems which provide company decision makers with the ability to connect "points of wisdom" about what needs to be done within and outside the organization for optimal results over time.

Relationship of Knowledge Management, Business Intelligence, and Optimization to Wisdom Management

As will be seen in the text, knowledge management provides a basic framework for business intelligence systems and optimization as found in smart business systems. In turn, both business intelligence and optimization help decision makers make better decisions. In addition, knowledge management is related to wisdom management in that wisdom is the umbrella that brings together not only knowledge in the form of business intelligence and optimization, but also provides the means to assist decision makers in reaching optimal and wise decisions over time. From this view, a new type of systems, that is, optimal KM/WM systems, can be defined. Such systems are forward looking by utilizing creativity and problem finding to the highest degree possible. They utilize the latest new business models that are integrated with e-commerce and the Internet. Overall, optimal KM/WM systems draw upon the basic concepts found in knowledge management, business intelligence, and optimization. As noted earlier, these newer type systems provide a company's decision makers with the ability to connect "points of wisdom" so that optimal decisions regarding forthcoming opportunities and upcoming problems can be reached today and into the future.

Concepts and Applications of Optimal Knowledge Management/ Wisdom Management Systems

The initial focus in the text is on important concepts underlying optimal KM/WM systems and the need to grow the learning organization over time. The integration of creativity with problem finding as they relate to these newer systems is examined. The development of new business models that facilitate the ability to judge soundly will be explored. Where applicable, newer techniques that focus on a *holistic* approach to an organization's opportunities and problems will be presented. Among these are software packages that center on optimization, goal programming, product lifecycle management, predictive analytics, knowledge discovery (data mining), data visualization, and virtual reality. Throughout the text, there will be a continual reference to getting the big picture in the areas of corporate planning, marketing, finance, and manufacturing. Additionally, the employment of a learning organization to adjust organization operations to meet changing times will be treated. All in all, the essential concepts underlying the development and implementation of these systems are treated. In addition, these concepts form the basis for an extensive treatment of what is the future direction of information systems.

Essentially, the first half of the text centers on the underlying concepts of optimal KM/WM systems. The second half of the text, in contrast, examines essential materials needed to develop and implement these systems. The focus is on their *applications* to assist typical organizations whether they be small, medium, or large. This complete coverage of concepts and applications centering on optimal KM/WM systems is focused on making decision makers more productive. Such systems allow decision makers to connect "points of wisdom" in new and different ways not found in the past for judging more soundly about their decisions about what needs to be done over time.

Text is Useful to a Wide Range of Professionals

The text is designed to assist a wide range of professionals. Top management and their staff, including the board of directors, will be particularly interested in getting involved in developing and implementing optimal KM/WM systems over time. Managers at all levels in the functional areas of a typical organization can also benefit from this text. Information system professionals will find the text helpful in understanding the tie-in of past information systems with optimal KM/WM systems. Consultants will find the text suitable for assisting their clients in moving to this new operating mode. Additionally, the text is suitable for academicians since it can be used in undergraduate and graduate courses. Overall, the text is suitable for anyone desiring to move to a higher level of systems operations on a day-to-day basis that really optimize an organization's performance over time.

Structure of Text

The text's structure is a logical one for a robust treatment of optimal KM/WM systems. The topical areas which are applied to the real world where appropriate are as follows:

Section I: The Challenge of Optimal Knowledge Management/Wisdom Management Systems

The focus of **Chapter I** is on the emergence of wisdom and its related activities to assist decision makers in getting the "big picture" and managing organizational operations on this basis. Optimal knowledge management/wisdom management systems stress the importance of connecting "points of wisdom" that are not found in past or current information systems of various types. Initially, the chapter focuses on answering the question — "what needs to be done" to optimize the operation of a learning organization. In turn, there is a discussion about information and its tie-in with their upper levels — wisdom and truth. Not only is there an introduction to optimal KM/WM systems, but also there is a rich discussion centering on the elements underlying them. This background serves as a basis for defining these newer systems. Also, there is an introduction to functional areas found in optimal KM/WM systems along with two typical applications of these systems.

Section II: Underlying Concepts of Optimal Knowledge Management/Wisdom Management Systems

In **Chapter II**, the basics of creativity are set forth, followed by the utilization of problem finding where the main thrust is on being "proactive" within an optimal KM/WM system-operating mode. Within this framework, an organization that encourages an acceptable tolerance for failure that come from experimentation of new opportunities and resolving future problems is on the right track for undertaking the development and implementation of optimal KM/WM systems. This new mind set is required to meet the challenges facing typical decision makers of today. A fundamental shift is thinking, especially from a creative thinking and problem-finding viewpoint, is needed to beat or, at least, meet global competition today and in the future. As such, this new orientation means that creative thinking and problem finding need to be an integral part of an organization's corporate philosophy for optimal decision making.

Chapter III covers the basics of computer storage and networking for optimal KM/WM systems. At the outset, future computer technology that will assist in developing optimal KM/WM systems is explored. In addition, business process management and information lifecycle management are examined. Effective computer storage follows next that centers on storage of aged data along with the need for a data federation approach for real-time data. The types of local and corporate wide databases are examined. In the second part of the chapter, the focus is on computer networking that includes wired and wireless technologies. There are a number of topical areas covered, including the Internet and the World Wide Web along with e-commerce. Typically, networking operations must be managed with greater levels of reliability and security than in the past.

In the pursuit of judging soundly about connecting "points of wisdom," **Chapter IV** explores a wide range of current software packages that are helpful in implementing and growing optimal KM/WM systems. The software explored includes the following: new business models, optimization, goal programming, product lifecycle management, predictive analytics, and knowledge discovery (data mining). In addition, data visulation software and visual reality software are included. Still other software packages could have been included, such as business intelligence and online analytical processing. In the near future, it is expected that newer software packages will be developed that truly fit under the category of optimal KM/WM systems. In cases where complete optimization is not practical, as in poorly-structured problems, near optimum solutions using the previously mentioned software packages are beneficial to a company's decision makers.

Section III: Building Optimal Knowledge Management/Wisdom Management Systems

In **Chapter V**, the design and implementation of an optimal KM/WM system requires taking an enlightened approach to the whole development process. That starts with applying creativity to the whole design process that is covered initially in this chapter, followed by the application of the KISS (keep it simple, straightforward) principle. In turn, EAI (Enterprise Application Integration) provides a framework for designing optimal KM/WM systems. Within this design framework, "points of wisdom" can be connected for producing optimal results. Next, the steps essential to developing and implementing effective optimal KM/WM systems are set forth. Included in these steps are those that relate to developing initial applications for connecting points of wisdom and disseminating results to grow wisdom of decision makers over time. In addition, there is a need to transform wisdom into action to meet decision maker's needs over time.

Section IV: Applications of Optimal Knowledge Management/Wisdom Management Systems

Chapter VI on corporate planning looks at the need to reinvent the organization for optimal decision making, which is influenced by a number of management principles. The important elements necessary for the development of an effective corporate planning model within an optimal KM/WM system environment are set forth along with the model itself and its sub-models. Next, short- to long-range corporate planning is tied-in with executive visioning, problem finding, venture analysis modeling, and evaluating corporate performance. Also, an optimal KM/WM system application that centers on corporate planning is illustrated for a holistic approach to an organization's operations.

To meet the challenges of the 21st century, **Chapter VII** explores the new marketing power of today's customers, especially as found on the Internet. Next, an enlarged view of connecting "points of wisdom" in marketing is set forth in a number of management principles not found in the past. The essentials necessary to develop an effective marketing model for optimal KM/WM systems are presented along with an appropriate marketing model and its sub-models. Although many marketing areas could have been explored, the focus is on marketing strategy and sales plans that connect marketing strategy to "points of wisdom" for judging soundly as well as making this strategy an integral part of venture analysis modeling. In the last part of the chapter, an optimal KM/WM system application in marketing is given.

At the outset, **Chapter VIII** examines an organization's visioning and its challenging financial goals. The relationships of globalization to optimal financial decision making and the need to take a global financial and accounting viewpoint are studied. An enlarged view of connecting "points of wisdom" in finance are set forth in a number of newer management principles. Next, the essential elements that underlie an effective finance model for optimal KM/WM systems are examined. In turn, a finance model and its sub-models found in optimal KM/WM systems are set forth. Because the text centers on making effective or wise financial decisions today and tomorrow, the area of financial performance is examined in some depth. In addition, an optimal KM/WM system in finance is given that takes a holistic approach to a firm that is experiencing growing financial problems.

The first part of the **Chapter IX** looks at the need to rethink manufacturing operations from a broad perspective so that optimal decision making in this area is the order of the day. Additionally, conventional wisdom versus an enlarged view of connecting "points of wisdom" in manufacturing are examined. The requirements for developing an effective manufacturing model for an optimal KM/WM system are explored, followed by a manufacturing model and its submodels. Due to the importance of production planning and execution in determining what should be produced daily, it is examined in detail with particular emphasis on making it an integrated part of product lifecycle management. Also, an optimal KM/WM system application is given for manufacturing.

In **Chapter X**, there is a last look at creative thinking and problem finding. Many times, a creative approach may signal the need for new business models that are more reflective of the times. Next, a review of appropriate computer software that is useful in optimal KM/WM systems is set forth. The main focus of the chapter is on effective applications in the areas of corporate planning, marketing, finance, and manufacturing. Additionally, an application is given as a holistic approach at the end of the chapter. For all companies, the accent is on assisting organizational personnel at all levels on a day-to-day basis for connecting "points of wisdom" within and outside the organization for optimal decision making over time.

Section V: The Impact of the Future on Optimal Knowledge Management/Wisdom Management Systems

Chapter XI examines a fourth-dimensional view that underlies newer developments that really affect optimal KM/WM systems. Future computer storage, networking, and software developments are discussed, followed by important considerations for developing and implementing a well-designed system. Future developments in the areas of corporate planning, marketing, finance, and manufacturing are examined and their tie-in with improving a company's decision makers wisdom. Also, there is a discussion on the continuing need for supporting optimal KM/WM systems to assist a company's decision makers over time. Lastly, there is a concluding reference to truth management systems.

Acknowledgments

We wish to thank the many professionals who have graciously assisted us. First, we want to thank the many management and information systems professionals who have supplied materials directly or indirectly for inclusion in this text. Second, our graduate students over the years at Xavier University (in Cincinnati, Ohio) are to be commended for their helpful suggestions and real-world experiences. Typically, these students, who are employed full time by a wide range of organizations in various industries throughout the Midwest, are management, marketing, accounting, manufacturing, and computer professionals. Third, a special note of appreciation goes to IGI's development editor, Ms. Kristin Roth, for her comments, as well as those of the reviewers on this very important subject that greatly affects the direction of the typical organization today. All in all, this publication would not have been possible without the assistance of theses professionals.

Section I:

The Challenge of Optimal Knowledge Management/Wisdom Management Systems

Chapter I

Introduction to Optimal KM/WM Systems

Issues

- To explore the importance of the question, *what needs to be done*, for today and tomorrow
- To examine the difference between conventional wisdom versus a wider view of wisdom
- To take a look at the "big picture" when taking a holistic approach within an optimal KM/WM system environment
- To examine two successful applications of optimal KM/WM systems as found in organizations today

Introduction

Today, how well an organization can respond to changing times is paramount. Decision makers at all levels need to adjust to meet fast changing times in a typical company. Basically, a company needs to change its information system to fit the times. In turn, appropriate computer techniques and technologies can be applied that best meets the requirements for the changed business conditions. The current failures of small- to large-sized organizations indicate that their information systems are not reflective of current business conditions even though the application of newer techniques and technologies may abound in the company. The turbulence of current business conditions, then, necessitates the need for decision makers to use the latest in information system developments, that is, optimal knowledge management (KM)/Wisdom Management (WM) systems. Initially, the chapter focuses on answering the question -"what needs to be done" to optimize the operation of a learning organization. In turn, there is a discussion about information and its tie-in with their upper levels - wisdom and truth. Not only is there an introduction to optimal KM/WM systems, but also there is a rich discussion centering on the elements underlying them. This background serves as a basis for defining these newer systems. Also, there is an introduction to functional areas found in optimal KM/WM systems along with two typical applications of these systems.

Focus on Answering the Question "What Needs to be Done?"

To avoid past mistakes of a typical organization, it is helpful at the outset to state that company decision makers tend to ask the same questions which most people ask. In turn, these decision makers assume that these are the right ones to answer. The better approach is not to give answers, but to ask questions of importance to the organization for growth opportunities. To state it another way, there should be a willingness by company decision makers to start not with the question "What do you think you should do?", but with the question, "What needs to be done" to develop and grow an organization that is committed to optimize its performance today and tomorrow. The willingness to change the way questions are asked can have a profound impact on making wise judgments about growing an optimized organization. Succeeding in a changing environment may mean abandoning what has worked in the past. This is the basic approach found in optimal KM/WM systems. Asking and answering the right questions for the times is at the center of judging wisely by a company's decision makers. Related to asking the right questions and getting wise answers center on the "big picture," which may mean reinventing the organization. That is, there is a need for a tie-in of executive visioning with corporate planning for the big picture about what needs to be done from the short to long term for growing an optimized organization.

As will be seen throughout the text, a number of newer techniques and software packages that focus on a holistic approach to resolving an organization's problems and capitalizing on its opportunities will be presented. Among these are:

(1) new business models, (2) business planning, (3) optimization, (4) goal programming, (5) product lifecycle management, (6) predictive analytics, and (7) balanced scorecard. For example, product lifecycle management (PLM) takes an enlarged view of a manufacturer's products and related services that addresses and incorporates the strategic business objectives and goals of the company. The same approach can be applied to a service-oriented company. Product lifecycle management provides an underlying framework for collaborating and sharing of product information, knowledge, intelligence, and optimization results among strategic partners, both within and outside the company's organizational boundaries. This PLM approach occurs over the course of a product or service's life from initial conception to taking it off the market. As such, PLM is business driven that is assisted by appropriate technologies. The successful manufacturer or service provider of tomorrow will be those that can provide their customers with the best products or services over their lifecycles that meet specific needs at the most competitive prices. Due to the importance of PLM, its essentials will be integrated throughout the text and related to venture analysis modeling.

Relationships Among Information, Knowledge, Intelligence, and Optimization and Their Systems

At the outset, it is helpful to examine the relationships among information, knowledge, intelligence, and optimization levels — refer to Figure 1.1 — and their systems. Underlying these levels is *data* that represents the unstructured facts and figures. Basically, it is the "data soup" of information systems at the very lowest level — widespread at the start of information systems development. In the early part of the 20th Century, Albert Einstein stated that he never wastes memory on things that can be easily stored and retrieved from elsewhere. For the most part these early data processing systems were batch oriented. Initially, they were manual systems that then moved to punched card equipment and finally to computer systems where the focus was on making them integrated computer systems. At that time, decision makers had moved well beyond data tabulation to the basic information level (first level on Table 1). Basically, information is structured data that is useful to decision makers in analyzing and resolving critical company problems. Companies found that they actually lowered costs, increased profits, and enhanced their market image using the latest information systems (refer to Table 1). These systems included real-time (Thierauf, 1975), distributed (Thierauf, 1978) decision support (Thierauf, 1982,

Table 1. Relationship of information to wisdom (and truth)

Level of Importance	Definition	Problem Importance	Type of System
Truth	Conformance to fact or reality	Vital	Not defined at this time
Wisdom	Ability to judge soundly over time	Critical	Wisdom management system
Optimization	Monitor operations for best solution	Major	Smart business system
Intelligence	A keen insight into understanding important relationships	Extremely broad	Business intelligence system
Knowledge	Obtained from experts based on actual experience	Major	Knowledge management and expert systems
Information	Structured *data* useful for analysis and decision making	Major to minor	Real-time, distributed, decision support, executive, and on-line analytical systems

1988, 1989), executive (Thierauf, 1991), and online analytical processing (OLAP) systems (Thierauf, 1997).

Knowledge, at the next level in Table 1, is obtained from experts based upon actual experience. As such, there is need to integrate a range of information in order to see patterns and trends that enable a decision maker to make the transition to insight and prediction. Essentially, this is the function of broad-based knowledge management systems (Thierauf, 1999) that go beyond expert systems (Thierauf, 1990) as found in the past.

At the next higher level per Table 1, there is the ability to understand the interrelationships of presented facts — whether they be information and/or knowledge — in such a way to guide action toward one or more desired goals. For the most part, intelligence centers on insight and understanding of the area, opportunity, or problem under study. It should provide the decision maker with the capability of meeting most situations whether they are unstructured or semi-structured. Such a capability is found in business intelligence systems (Thierauf, 2001).

Even though business intelligence has helped a wider range of individuals within and outside the typical organization, the next level per Table 1 is a move toward optimization that assists these same decision makers to allocate a company's resources in a more effective way. Basically, optimization means that after all information and knowledge have been thoroughly understood (i.e., intelligence), the next level of importance to decision makers is the optimization of resources that they have at their command. Today, optimization is found in smart business systems (Thierauf & Hoctor, 2003). Effective smart business systems are concerned with monitoring a company's operations that result in optimization or near to optimization as possible. As such, smart business systems can enhance the effectiveness of a company's operations by providing decision makers with very desirable solutions.

Fundamentals of Wisdom

Wisdom, shown at the second highest level per Table 1, is the ability to judge soundly about what needs to be done to grow a learning organization over time. This high level of understanding involves such philosophical attributes as the awareness that the models constructed will not always hold true. Hence, a beginning point for wisdom can be a humble assertion: "I don't know." This assertion can be the real beginning of wisdom. Because business transactions per se do not inspire much wisdom for decision makers, wisdom grows from making the connections of these transactions to each other and their change over time. Wisdom requires an intuitive ability, born of experience, to look beyond the apparent situation in order to recognize exceptional factors and anticipate unusual outcomes. Wisdom allows decision makers to visualize opportunities or see problems in a new light in order to cut to the heart of the relationships for what needs to be done. Essentially, wisdom is a vital organizational resource, accumulated through experience, and applied to everyday learning at work. Because wisdom is a personal capacity acquired through experience and thinking, past hierarchical and functional roles can be replaced with learning relationships that focus on wisdom as the foundation of the new organization. Wisdom can be used as an organizational strategy to realize the real potential of any organization.

Linking of Knowledge to Wisdom

As noted above, knowledge provides a framework for intelligence and optimization. Basically, intelligence centers on helping decision makers gain insight and understanding of an opportunity or a problem under study. On the other hand, optimization focuses on monitoring a company's operation to assist decision makers in reaching best or near best solutions. For both intelligence and optimization, knowledge is an integral part of their essential elements for assisting decision makers. Going one step further, wisdom is the umbrella that brings together not only knowledge in the form of intelligence and optimization, but also provides the means to assist decision makers in reaching wise and optimal decisions over time.

Typically, a wise decision maker knows what knowledge in the form of intelligence and optimization is needed in a given situation to optimize an organization's operations for what needs to be done over time. The individual knows how to engage in problem finding for developing important organizational opportunities and solving organizational problems. Hence, there is a tie in of knowledge and wisdom that is recognized in the subject matter of the text - optimal knowledge management/wisdom management systems. Optimal KM/WM systems are the means for assisting decision makers at all levels of the organization as well as its customers and trading partners to reach optimal decisions over time.

What Lies Beyond Wisdom?

At the highest level per Table 1 is truth, which represents conformance to fact or reality and represents the lofty pinnacle of understanding. It comes from understanding the way that connected points of wisdom come together. Although its place in the typical organization is being debated at this time, it is safe to say that certain truths centering on ethical and environmental issues are always useful to the typical manager for guiding a company's direction at all times. A violation of basic truths held by the general public can only jeopardize its standing in the community. Going beyond truth as found in the business community is that found in the religious community. To the ultimate degree, truth is equivalent to God. At this time, they can be called truth management systems even though their essentials have not been defined at this time.

Introduction to Optimal KM/WM Systems

Optimal KM/WM systems will make use of newer computer technology to take advantage of the vast amount of data, information, knowledge, intelligence, and optimization storage resources available to decision makers for what needs to be done organizationally. These type systems can be looked upon as an expansion of knowledge management systems, which include business intelligence systems and smart business systems that inspire the loyalty and trust of a company's customers and its employees. A comment from Aristotle — one of the great masters of the past — is relevant here: "Action without knowledge is folly. Knowledge without wisdom is perilous." Company managers operating within an optimal KM/WM system environment will encourage customers and employees to generate ideas for new products and services. Similarly, wise managers will assist their customers and employees to explore promising opportunities and to anticipate problems and solve them before they happen as well as helping both achieve their objectives and goals such that there is a win-win situation. Wise managers will also help both identify opportunities and problems for solutions that come out of problem finding.

Conventional Wisdom vs. an Enlarged View for Connecting "Points of Wisdom"

Optimal KM/WM systems will have the capability of assisting decision makers in new thinking about their companies and themselves. The conventional wisdom, such as "If it ain't broke, don't fix it," will facilitate decision makers in their new thinking, such as, "There's gotta be a better way." Another bit of conventional wisdom states that there is a time and a place for everything. This new type system will help decision makers redefine the space/time continuum. Still another conventional wisdom item is that you get what you pay for. A newer direction from an optimal KM/WM system might be that one particular combination gives customers a lot more for their money (i.e., a real bargain, all things considered).

For a company to experience an enlarged view of wisdom, a company's decision makers must pursue continuous innovation strategies. The reason is that the value of a company's products and services can be retained for only a limited time. Some companies, by their very nature, must live by continuous innovation. Continuous innovation strategies often depend upon fast and often discontinuous learning. Such strategies are also related to value satisficing, that is, there is need to sacrifice today's value for future value in terms of profits. An enlarged view of wisdom is required to know when to move from value sacrifice in the short term to value retention over the longer term. Generally, an enlarged view of wisdom can be enhanced through the extensive use of business analytics and related predictions by providing decision makers with the ability to recognize when old rules, procedures, and the like no longer apply and that new ones need to be created.

To assist decision makers in getting an enlarged view of wisdom, an emphasis in this text is connecting "points of wisdom" such that decision makers can make even more sound judgments about areas under their control for what needs to be done to grow the learning organization. Decision makers can take a comprehensive approach to operations under their control today and tomorrow. A recommended approach to connecting points of wisdom is the employment of workflow systems, which handle processes involving human input. Today, this is known as business process management (BPM). Basically, BPM allows organizations to analyze workflow and best practices to manufacture, market, deliver, and support their products and services in the most efficient manner possible for better bottom-line performance. Through BPM, organizations can streamline, automate, and automate business processes to continuously improve and enhance business performance.

Utilizing business process management, decision makers can determine where points of wisdom need to be connected for optimizing an organization's operations for what needs to be done over time. For example, Sam Walton who founded Wal-Mart recognized the force behind the supply chain as it is currently practiced. He brought about this shift in market savvy through his accent on customers. Walton had taken a hard, that is, wise, look at the customer, and saw quite clearly that the actions of the customers in his stores were the keys to expanded business. He helped create a system whereby, as a product left the shelf through customer choice, the store instantly monitored the sale and was able to initiate a replacement of the product automatically. In effect, he connected in a most important way how the customer and the supplier interact, that is, he connected the points of wisdom such that his stores became the enabler of this process. Throughout the process, Sam Walton focused on the big picture for a comprehensive approach to the company's operations.

Consider Global Forces, but Act on a Local Basis

When a company's decision makers connect "points of wisdom," it is recommended that they take a holistic approach for linking operations around the world. That is, the global forces facing the business community whether they be political, economic, social, technological, human, or regulatory need to be addressed and acted upon by a company's decision makers. In turn, these forces determine where U.S. manufacturers (and service companies) will be experiencing growth as contractors. Currently, there are a number of countries that use hot spots for design and production as well as key markets that U.S. manufacturers are targeting for globalized production. Even though China has been getting a lot of attention lately, there are closer destinations, that is, Canada and Mexico, that U.S. manufacturers are entering for design, production, and even for market growth. In Europe, U.S. manufacturers are investing heavily in established countries, like the U.K., France, and Ireland. Additionally, they are rapidly increasing investments in lower cost countries such as Czech Republic, Hungary, and Russia. The question today is "where" should a company globalize (Moad, 2005).

As expected, there will be winners and losers, which are the result of global forces. International companies must take advantage of these global forces so that there is a win-win situation. For example, Procter & Gamble provides products to 140 countries around the world. Hence, it must think globally and stabilize global forces to its advantage. In turn, it must act on a local basis so that everyone, including its customers, trading partners, employees and local community, benefit from the company operations. It should be noted that Procter & Gamble is not outsourcing just for cost savings but also to target new markets and to leverage regional efficiencies.

Consider the Impact of Disruptive Technologies

An important factor affecting optimal KM/WM systems is the area of disruptive changes not only globally, but also locally. Disruptive changes can be caused not only by a number of factors that are related to a company's markets, but also can be related to a company's size per Clayton Christensen. When a company is young, its personnel, equipment, technologies, brands, and the like, define what it can and cannot do. As it becomes more mature, its abilities, for example, stem more from its processes, such as product development, improved manufacturing, and financial capabilities. Because companies, independent of the people within them, have capabilities, those capabilities also define disabilities. What a company can and cannot do as it grows becomes more clearly defined in certain predictable ways. In the largest companies, values, particularly those that determine what are its acceptable gross margins and how big an opportunity has to be before it becomes interesting, define what the company can and cannot do. In contrast, smaller companies tend to respond to major market shifts better than larger ones because resources are more adaptable to change than processes or values.

To better understand disruptive changes, it is helpful to look at the past. In the early 20th Century, for example, one had to be a J.P. Morgan-type to buy stocks. Charles E. Merrill's strategy, on the other hand, was to bring Wall Street to Main Street. He did this by creating a business model of salespeople who were not commissioned but were salaried so they could afford to sell stocks to ordinary Americans. By making it so simple and relatively cheap, he brought a larger population into the investing market. Now, Charles Schwab and E-Trade have done it again, enabling even college students to be day traders and manage their own portfolios. In a similar manner, Canon and Ricoh displaced Xerox by making copiers so low in cost that people could do their own copying (Christensen & Overdorf, 2000).

The Essential Elements Underlying Developing and Implementing Optimal KM/WM Systems

The essential elements underlying developing and implementing optimal KM/WM systems center on those that assist decision makers in making the shift to judging their decisions more soundly on a day-to-day basis. Basically, the focus is on viewing their organizational operations in a manner that was not possible before. To accomplish this modus operandi, the five essential elements found in these newer systems are:

1. Focus on the big picture of organizational operations for "what needs to be done" over time
2. Undertake creative thinking as related to problem finding
3. Concentrate on those activities that create and distribute unique values for all parties
4. Employ newer business models and computer software and techniques for developing new opportunities and solving organization problems
5. Use a learning organization to adjust to changing times

All of these essential elements are noted in the next sections and are covered in future chapters of the text.

An integral part of these five elements is trust that is critical to an organization's effectiveness. Although a decision maker is trustworthy does not guarantee that the individual is capable of building trust in an organization. Traditional managerial virtues like consistency, clear communication, and a willingness to tackle difficult questions must be related to potential enemies of trust. Among the most common enemies of trust are inconsistent messages from top management, inconsistent standards, a willingness to tolerate incompetence or bad behavior, dishonest feedback, a failure to trust others to do good work, a tendency to ignore painful or politically charged situations, consistent corporate underperformances, and rumors. Fending off these enemies of trust must be at the top of every chief executive's agenda. But even with constant vigilance, an organization and its leaders will sometimes lose people's trust. During a crisis, it is recommended that managers enlist the help of an objective third party for best results. If a company's managers "go dark" in the face of a crisis, employees will worry about the company's survival, about their own capacity to cope, and about the managers' ability as leaders. Thus, trust is necessary before optimal KM/WM systems can be developed and implemented (Galford & Seibold Drapeau, 2003).

Focus on the Big Picture of Organizational Operations for "What Needs to be Done" Over Time

An important part of developing and implementing optimal KM/WM systems is centering on the big picture for a holistic approach for "what needs to be done" to grow the learning organization over time. As a starting point, a company's decision makers need to expand the company's customer base and their retention over the long term. Today, there is generally need not only to design and build customized orders for customers, but also to lower costs in the process of doing so for all parties involved in the product or service. Next, progressive companies must accelerate the time-to-market, that is, they must shorten the product design cycle as well as optimize procurement from all sources of supply. Additionally, there is need to accelerate time-to-market by increasing volume over a short time period. Needless to say, supply-chain relationships need to be strengthened for this to happen. The bottom line of this holistic approach to effective organizational operations is increased profitability for all stakeholders - a company's customers, the company itself, its suppliers, its employees, and its financial resources from within and outside the organization. Profitability can be accomplished in a number of ways, such as optimizing interactions among stakeholders and automating routine tasks. Overall, the utilization of a holistic approach centers on making an organization more profitable over the long term. This approach will be found throughout the text, especially in Chapters VI through X.

As indicated, the big picture today for what needs to be done centers on extending beyond the confines of the organization and include all of stakeholders along with those elements that have direct or indirect impact on the organization. This includes not only the Internet, but also other elements. One such important element is the movies which most of the general populace go to see on a regular or a periodic basis. For example, any company that sells products or services to the general public on a continuous basis, such as Coca-Cola, are well advised to make sure that the product or service appears or is mentioned in the movies. In this way, a person who sees these products being used or consumed in the movies will consciously or subconsciously pick up these products or use these services when shopping.

Undertake Creative Thinking as Related to Problem Finding

The whole range of creativity and its relationship to problem finding as found in optimal KM/WM systems will be covered in the next chapter. At this point, it is safe to say that a creative approach starts with an organization's customers. Such an approach can be stated in series of questions. How many new ideas do a company's employees generate that center on the customer? How well does the company carry them out? Which of the company's traditions, policies, and rules get in the way of fulfilling customer's needs? Who are the company's most creative people? How does the company go about hiring them? In turn, a series of questions can be generated that relate to manufacturing operations, costing methods, human resource relations, trading partners, and the like. Needless to say, creative people need to be nurtured in the workplace so that a company's creative assets (i.e., its intellectual capital) will more than pay for themselves over the long term for an organization's total operations. It is counterproductive to talk about creativity and then retain a working environment that deadens the imaginative and the creative spirit of decision makers throughout a company.

Concerning the need for problem finding, the authors and other management experts saw this a long time ago. Because most information systems have been and still are, to a degree, built on accounting data and information, they tend to be inwardly focused. If led by these systems, decision makers are consumed with what they already know about, such as sales and costs. In contrast, decision makers need to focus on what they find more difficult: the creation of value and wealth. Peter Drucker's related belief is that newer system approaches must provide decision makers with more pertinent external and internal information, knowledge, intelligence, and optimization results. But that which is created by current information systems does not seem to be helping decision makers make the really tough and important decisions. What is needed are information systems that are capable of providing decision makers with much more than queries, report generation, and warehousing. Rather newer information systems are the culmination of major system technologies that provide the ability to take a proactive stance concerning what needs to be done organizationally rather than a passive or, at best, a reactive approach to a company's operations. A proactive approach, that is, a means for identifying an organization's opportunities and future problems that need to be solved today, is at the center of optimal KM/WM systems.

Creative thinking, as related to problem finding, needs to focus on a new paradigm, that is, a new way of viewing the company's customers. An excellent example is the utilization of the Internet by a company and its customers in terms of selling automobiles. More specifically, a typical customer uses the Internet not so much to buy a car, but rather to eliminate certain cars from consideration when buying. Thus, when a potential customer comes to a specific car deal, the individual knows what he or she does not want to buy. From this perspective, the Internet represents a new sales distribution channel for potential customers. The Internet is essentially a buyer for the company's customers and not the traditional seller. That is, the automobile dealer finds the right car for the customer, including competing cars and the right dealer where the car can be found today. The bottom line in this creative approach to the buying and selling of cars is that the customer can save valuable time, not to mention money by acquiring a vehicle that meets the individual's needs. The traditional role of the automobile dealer has been changed by now being focused on meeting a customer's specific car needs rather than trying to sell a customer a car that he or she may or may not want.

Concentrate on Those Activities that Create and Distribute Unique Values for All Parties

In the process of defining the customer, there may be need to rethink a company's business. Related to this rethinking process is incorporating activities that create and distribute unique values for customers. A company's value proposition means spelling out how a company's business is different from all other competing firms, that is, its unique value proposition. The product or service that the company provides should be recognized as different and what the customers are willing to pay for it. Essentially, this unique value proposition addresses a specific market need. This includes identifying the unmet or latent needs of a large market. This is what Southwest Airlines did. It owes its success to identifying the unserved market before innovating its travel solution for its customers.

Once the target market has been segmented, there is a paramount need to inform customers how its products or services are different from others. Fundamentally, this centers on spelling out the company's unique value proposition, that is, the focus is on creating and distributing unique values for customers, suppliers, and employees alike such that there is a win-win situation for all involved. For customers, this product and/or service will help them improve performance, assist in problem solving and problem finding, reduce operating costs and risks, and improve their quality of life. For suppliers, this unique value proposition may take the form of allowing access to research and development facilities and reaping the benefits of centralized buying to benefit from economies of scale. For employees, the focus of unique value proposition could be on assisting them in their career goals, providing training to grow their knowledge and skills, and providing a means to meet their financial retirement goals. Overall, a unique value proposition needs to be articulated to all parties within and outside the organization for best results (Sandberg, 2002).

At the core of the value proposition for a new product and/or service is the need to answer a series of questions. Will the company build its own plant or outsource all or most of the manufacturing operations? What distribution networks will the company need to generate requests for the product or service the company intends to provide? How will the company ensure that the people providing the actual service have the proper training? Needless to say, many other questions need to be addressed by the company's management. Typically, creating and distributing in accordance with a company's value proposition involve collaborating with other companies. Sometimes, the collaboration is horizontal, as when several of the major automakers created a joint initiative that enabled them to purchase supplies more cheaply. A clothing manufacturer that enters into an alliance with a retail store chain to gain better access to customers at the point of sale is an example of vertical collaboration. Overall, an organization's value proposition should weigh the myriad alternatives in terms of costs and tradeoffs involved with such collaborative strategies and describe the optimal configuration of value chains and networks. The concept of value proposition will be noted throughout the text, particularly in Chapter VII on marketing.

Employ Newer Business Models and Computer Software and Techniques for Developing New Opportunities and Solving Organizational Problems

Today, the employment of newer business models and computer software and techniques to facilitate wisdom at the various levels of management to develop new opportunities and solve organizational problems is paramount. In terms of business models, conventional wisdom of the past focused on companies with the largest market shares who, in turn, would have the largest revenues and lowest costs for the highest profits. However, many large companies, like General Motors and IBM, have found this not to be true. What is happening is that companies are seeing a migration from products and services to what might be called "spinouts." One of the newer business models for organizations for changing times is centering on spinouts as "drivers of profitability." For example, General Electric's jet engine profits are not in building the engines, but in their financing, service, spare parts, and overhauling existing engines. Basically, manufacturing of these engines has become almost incidental to the total operations for producing profits. The essential message here is that a typical organization must change its basic organization business model to reflect changing times. Typically, organizations should upgrade or change their business models periodically. They should have other models in reserve in case the current one becomes obsolete.

A most important factor affecting business models are economic downturns that present important opportunities to strengthen a company's relative strategic position. Essentially, the business model should protect a company's best people, reinforcing relationships with its best customers and suppliers, and distributing new values for all in the value chain. In a downturn, top-level decision makers make the mistake of cutting costs across the board instead of getting rid of lowest-value producers and the weakest performers. Whether intentional or unintentional, this indiscriminate cutting can kill off the very business model that is growing into the next profit engine for a company.

Another situation calling for a change to an organization's business model is when a new product does not fit in with the old paradigm. Two examples come to mind – IBM's approach to the PC market and Firestone's approach to the radial tire market. IBM's existing business model centered on selling computer mainframes, which did not allow it to imagine a market for PCs. Firestone's business model, geared for making bias tires, could not be adjusted to manufacture radials. At the outset, both companies failed to incorporate product innovation in their business model. Still another situation calling for changes to an organization's business model is an existing product that has entered a new stage in its product life cycle. The business model needed to launch an innovative product differs dramatically from the one needed to meet high demand once the product catches on. Such is the case with innovative toys and sport products for children. Top-level decision makers should make the current organization business model adaptable to a product's passage throughout its life cycle. Newer business models for what needs to be done to grow the optimized organization over time will be noted where deemed appropriate in the text.

Related to newer business models are computer software and techniques. More advanced software technologies are needed since the business lifecycle for new products and services are getting shorter over time. In the process, there has been a fundamental shift in customer behavior and distribution of products and services. For example, a manufacturer of quality furniture would need to consider this shift in its current and future operations. For the current year, a computerized goal programming approach could be employed to achieve profit goals for the wide range of furniture lines being produced. From another perspective, product lifecycle management software could be used to determine the maximum profits over the lifecycles of its furniture lines for today and tomorrow. In turn, a wide range of financial analytics could be employed to obtain a thorough evaluation of the company's operations

The types of advanced computer software useful in optimal KM/WM systems along with their relevant techniques include the following: business planning, optimization, goal programming, product lifecycle management, predictive analytics, balanced scorecard, and knowledge discovery (or data mining). In addition, data visualization and virtual reality helpful in better visualizing the results. The main focus of these newer software packages and their related technologies is to optimize a company's resources within an optimal KM/WM system environment on a day-to-day basis. In cases where complete optimization is not possible as found in unstructured problems, it is possible for decision makers to make sound judgments using near optimal solutions. The subject matter of newer software and their related techniques that is conducive to assist decision makers in getting a good handle on present and future operations is given in Chapter IV.

Use a Learning Organization to Adjust to Changing Times

Within an optimal KM/WM system environment, a learning organization approach allows a company's decision makers to adjust to changing times. Essentially, a learning organization sweeps away structures and assumptions long ingrained within the top-down, bottom-line-driven organizations of the 20th Century. This newer direction will be a major driving force throughout the 21st century. Knowing what the company does well, learning from that to do it better the next time, and continually looking for improvement are the hallmarks of a learning organization. Although there is disagreement among the experts concerning its various aspects, most view a learning organization as one that develops over time and is linked with business intelligence and optimization results. In addition, its performance improves over time, which can be linked to improved financial performance. A learning organization is skilled at acquiring and utilizing facts and figures and, at the same time, at modifying its operations to reflect new patterns and insights. Peter Senge, who popularized learning organizations in his book, *The Fifth Discipline*, described them as places where people continually expand their capacity to create the results they truly desire, where new and expensive patterns of thinking are nurtured, where collective aspiration is set free, and where people are continually learning how to learn together (Senge, 1990).

Typically, leveraging intellectual capital (or assets) or providing learning on demand is an important trait of a learning organization. Treating the organization as a brain — a repository and processor of assorted know-how, information, and knowledge rather than a machine — deepens a respect for both the individual and the team as sources of a company's innovation. All this, of course, is part and parcel of a learning and intelligent company, and the stakes are larger than just meeting budgets and timetables. Today, a typical company uses approximately 20% of its intellectual capital. If, through a learning organization approach, a company can raise that to 30%, that represents a 50% gain, in some of the company's assets, say the experts.

A true learning organization assists organizational decision makers in creating values desired by the company's customers as well as a company's other stakeholders. A company's decision makers create an environment of inquiry and discovery for reaching desired values of all parties. Decision makers of learning organizations are adept at communicating a shared vision and helping others gain accurate views of reality. While a learning organization advocates leave no doubt as to the importance of the free expression of ideas, organization experts note that the exchange must be tempered for maximum efficiency. The value of intelligence and optimization results is in their use and not in their collection per se. That is, it is important not to collect and save everything but to relate what is known and not known to what needs to be known to fulfill organizational objectives and goals. In this manner, a learning organization can be leveraged for best results. The whole concept of the learning organization will be found throughout the text as well as in Chapter IX.

Optimal KM/WM Systems Defined

Basically, optimal KM/WM systems look at the organization in terms of the following elements (as discussed previously): (1) focus on the big picture of organizational operations for "what needs to be done" over time, (2) undertake creative thinking as related to problem finding, (3) concentrate on those activities that create and distributed unique values for all parties, (4) employ newer business models and computer software and techniques for developing new opportunities and solving organizational problems, and (5) use a learning organization to adjust to changing times. All of these elements are related to helping decision makers judge the soundness of their decisions when connecting the "points of wisdom" for what needs to be done to grow the learning organization over time. The bottom line is that decision makers should accumulate wisdom to undertake new opportunities and solve problems that cut through to the heart of the important relationships about what needs to be done.

Optimal KM/WM systems can be defined as systems for business that allow selected information, knowledge, intelligence, and optimization results be utilized by decision makers for connecting points of wisdom in order that sound judgment calls can be made to improve organizational operations for what needs to be done over time. They can involve the process of reinvention to meet changing times by focusing on the big picture for a holistic approach. These systems employ appropriate analytical and collaborative tools that operate within a global computer networking architecture. The type of newer business models and computer software and techniques used is situational. Typically, these systems provide decision makers with the ability to judge the soundness of their decisions over time in order to guide action for what needs to be done toward desired actionable objectives and goals over time. They provide decision makers with timely information, knowledge, intelligence, and optimization results in problem finding for improving a company's operations. Optimal KM/WM systems employ creativity to its fullest for creating and distributing unique values desired by all stakeholders in order to assist decision makers in judging the soundness of their decisions. Essentially, these systems cut through the heart of important relationships by connecting "points of wisdom" today and tomorrow about what needs to be done over time to grow a learning organization. Due to the expanded capabilities of optimal KM/WM systems over prior information systems, it is expected that they will be the mainstay of businesses for many years to come.

Functional Areas of Optimal KM/WM Systems

A logical starting point for developing and implementing optimal KM/WM systems is corporate planning which serves as an umbrella for the functional parts (i.e., marketing, finance, manufacturing, and human resources) of a typical company to determine what needs to be planned today and tomorrow. It is also referred in the literature as strategic planning and executive-level planning. The real essence of corporate planning is the capability of an organization to relate external sources to internal factors in such a manner that changes from external sources are used as input for making appropriate changes to the internal factors. This functional area is covered in some depth in Chapter VI.

The functional area of a typical organization that centers on creating and distributing unique values for customers is marketing. As such, a company's marketing efforts should lead customers where they will be in the future and allow them to get more involved in a company's operations from a learning perspective. To assist customers so there is a win-win situation, marketing activities start the flow of products and services from their inception to their final use by customers. From this broad-based perspective, marketing should anticipate the needs of customers (by being proactive) rather than follow what the customers need (by being reactive). This important area of marketing is set forth in Chapter VII.

The underlying means for bringing together marketing and related manufacturing activities is the functional area of finance (and accounting). Finance and its related accounting activities have come a long way today in terms of getting a real handle on a company's operations. Clearly, companies should outsource only when financial benefits can be quantified. Otherwise, a company should do most everything itself in order to improve the bottom line. Assessment of accurate financial information, knowledge, intelligence, and optimization results in the form of key performance indicators and financial ratios will go a long way toward assuring the short- to long-term viability of an organization. The area of finance and its related accounting activities is examined in Chapter VIII.

While marketing starts the flow of merchandise to customers, manufacturing provides the actual movement of manufactured goods — from the raw material state to work in process, to the finished goods state — to the desired destination by customers. Today, companies have taken a global perspective in terms of their manufacturing operations. Essentially, an organization would be well advised to do what its does best whether in the area manufacturing operations or otherwise. The focus needs to be on maintaining a manufacturing organization over time so that its costs are kept at a minimum. Manufacturing operations are covered in Chapter IX.

Although there is not a separate chapter or a separate section in each of the above chapters (Chapters VI to IX) for human resources, either approach could have been used. Fundamentally, the planning and implementation of corporate planning, marketing, finance, and manufacturing activities are undertaken by the human element who is assisted by computer technology. Similarly, the outputs are evaluated by company decision makers for changes needed to a company's operations to reflect the realities of the current times. The net result is a focus on what needs to be done for growing the learning organization over time.

Robust Applications of Optimal KM/WM Systems

Applications of optimal KM/WM systems seem to vary from industry to industry and companies within an industry. While some organizations have taken a comprehensive or a holistic approach to these systems, others have taken a segmented or functional approach before undertaking a comprehensive approach. An example of these two approaches are given next.

One celebrated example of a holistic approach operating within the framework of a newer organizational business model is Dell Computer. Dell's comprehensive holistic approach to its business model is centered on making it the industry's most efficient PC manufacturer. It uses a direct-sales model whereby Dell computers are made to order and then delivered to customers. Essentially, there is no middleman. The customers get the exact computers wanted that are cheaper than that from competition. Dell gets paid by its customer's weeks before it pays suppliers. As other PC vendors have struggled, Dell has thrived. Currently, Dell's latest organization business model is seeking to transform itself into a world-class e-business vendor. For Dell to be successful with this newer business model, it has to forge partnerships for its new e-services ventures. Dell recognizes that it cannot do everything by itself. Recently, Dell has bolstered its services and support offerings via a multibillion-dollar deal that calls for IBM

Global Services to provide on-site repair and warranty assistance for its customers worldwide. Dell also maintains service relationships with Unisys and Wang. While hardware will still be the heart of Dell's operations, its latest organization business model is centered on expanding the company's offerings to include a broad array of strategic platforms and to bundle those products with services that leverage its expertise in Web sites, supply chains, and Internet infrastructure maintenance. Overall, Dell Computer has become the benchmark for other computer manufacturers to emulate when reinventing the interworkings of their day-to-day operations for wise decision making for what needs to be done (Jones, 2003).

While Dell Computer has been very successful with a holistic approach, other companies have been successful using a segmented approach for optimal KM/WM systems. Procter & Gamble has been very successful to date with a segmented approach although there is a move toward a comprehensive approach. The company has focused its efforts on an effective supply-chain optimization system. The consumer goods giant has built its extensive supply chain on a global ERP (enterprise resource planning) system that unites operations in 130 countries and supports myriad products, from diapers to toothpaste to pet food. Another area where Procter & Gamble has been successful is in using information technology for product lifecycle management, that is, the art of creating and nurturing a brand, as noted earlier in the chapter. If companies like P&G do not get value out of all of their marketing efforts by creating new and profitable products, they lose their edge in the marketplace. Going another step forward, Procter & Gamble is seeking new and wiser ways to move its hundreds of brands through its global supply chain, creating a real-time business that is driven off product demand, not forecasts, and where products are delivered accurately, thereby eliminating manual intervention and costly product returns. That centers on using electronic catalogs, Web-based ordering systems, and radio-frequency ID chips designed to track individual products, among other technologies. On top of that, this giant consumer goods company seeks ways to create new product innovations that can make their way through sophisticated supply chain to land in front of consumers' eyes more quickly. Hence, the individual components of an optimal KM/WM system must be woven together for a comprehensive approach to what needs to be done over time such that there is a win-win situation for all stakeholders (Stahl & Soat, 2003). (This real-world example will be treated again in Chapter X.)

Other real-world examples of optimal KM/WM systems will be presented in Chapter X. Additionally, still other case studies are found in Chapters VI through IX. Although organizations are making great strides to implement an optimal KM/WM system-operating mode, they still need to take a holistic approach that reflects their ideal operating environment. To help the reader to better understand future developments that affect optimal KM/WM systems, Chapter XI details them. Essentially, these developments will further help decision makers to judge soundly about their decisions over time.

Summary

The turbulence of current business conditions that will undoubtedly intensify is related to answering the question: "What needs to be done?" Although the focus of past information systems was on information, knowledge, intelligence, and optimization, newer optimal KM/WM systems center on the ability of decision makers to judge soundly today and tomorrow about their decisions. An introduction to these systems was given where the conventional wisdom was compared to an enlarged view of wisdom. The focus was on connecting "points of wisdom" so that decision makers can make even more sound judgments about what needs to be done to grow the learning organization over time. Also, consideration for global forces where companies act locally as well as disruptive technologies were addressed. Next, the essential elements underlying developing and implementing optimal KM/WM systems were discussed at some length and formed the basis for defining them. Additionally, two robust applications of these systems were set forth. Overall, this first chapter sets the stage for the underlying concepts and practices of optimal KM/WM systems as they are found in business today.

References

Christensen, C. M., & Overdorf, M. (2000, March April). Meeting the challenge of disruptive change. *Harvard Business Review*, 67-76.

Galford, R., & Seibold Drapeau, X. (2003, February). The enemies of trust. *Harvard Business Review*, 88-95.

Jones, K. (2003, February). The Dell way. *Business 2.0*, 61-66.

Moad, J. (2005, July). Winning the globalization game. *Managing Automation*, 24-32.

Sandberg, K. D. (2002, January). Is it time to trade in your business model? *Harvard Management Update*, 1-4.

Senge, P. (1990). *The fifth discipline*. New York: Doubleday.

Stahl, S., & Soat, J. (2003, February 24). Feeding the pipeline. *Information Week*, 49-53.

Thierauf, R. J. (1975). *Systems analysis and design of real-time management information systems.* Englewood Cliffs, NJ: Prentice Hall.

Thierauf, R. J. (1978). *Distributed processing systems.* Englewood Cliffs, NJ: Prentice Hall.

Thierauf, R. J. (1982). *Decision support systems for effective planning and control: A case study approach.* Englewood Cliffs, NJ: Prentice Hall.

Thierauf, R. J. (1988). *User oriented decision support systems, accent on problem finding.* Englewood Cliffs, NJ: Prentice Hall.

Thierauf, R. J. (1989). *Group decision support systems for effective decision making: A guide for MIS professionals and end users.* Westport, CT: Quorum Books.

Thierauf, R. J. (1991). *Executive information systems: A guide for senior management and MIS professionals.* Westport, CT: Quorum Books.

Thierauf, R. J. (1997). *On-line analytical processing systems for business.* Westport, CT: Quorum Books.

Thierauf, R. J. (1999). *Knowledge management systems for business.* Westport, CT: Quorum Books.

Thierauf, R. J. (1990). *Expert systems in finance and accounting.* Westport, CT: Quorum Books.

Thierauf, R. J. (2001). *Effective business intelligence systems.* Westport, CT: Quorum Books.

Thierauf, R. J., & Hoctor, J. J. (2003). *Smart business systems for the optimized organization.* Westport, CT: Quorum Books.

Section II:

Underlying Concepts of Optimal Knowledge Management/Wisdom Management Systems

Chapter II

Creative Thinking and Problem Finding Underlie Optimal Decision Making

Issues

- To explore rethinking creative thinking in terms of what needs to be done organizationally over time
- To look at the various creative techniques that are useful to decision makers
- To explore the use of problem finding from the standpoint of turning problems into opportunities
- To examine how problem finding can assist in expanding the wisdom of decision makers

Introduction

In this 21st Century, there is an even great need for decision makers to ask the right questions to take advantage of future opportunities and solve upcoming problems. Within this framework, the chapter focuses on creative thinking and problem finding. An organization that encourages an acceptable tolerance for failure that come from experimentation of new opportunities and resolving future problems is on the right track for undertaking the development and implementation of optimal KM/WM systems. This new mind set is required to meet the challenges facing typical decision makers of today. A fundamental shift is thinking, especially from a creative thinking and problem-finding viewpoint, is needed to beat or, at least, meet global competition today and in the future. As such, this new orientation means that creative thinking and problem finding need to be an integral part of an organization's corporate philosophy for optimal decision making.

Rethinking Creative Thinking in Terms of "What Needs to be Done"

At the core of creative thinking as found in optimal KM/WM systems is the need to take a second look at creative thinking, that is, rethinking creative thinking. The focus is on "what needs to be done" versus "what do you think," as noted in the opening part of Chapter I. By doing so, decision makers then have the capability to reinvent the organization. It is not enough to create new products and services, but also there is a need to create a continuously innovating organization. By the same token, it is never enough to tell organization personnel about some new directions in creative thinking and problem finding. It is necessary to have them experience it in a way that gets them involved in a positive manner. Organizations need to encourage multiple points of view since interdisciplinary approaches can bring enormous value to problem finding and problem solving. In maximizing the differences in backgrounds, cultures, age, and like items, there is the increased likelihood that the result will not be what was originally imagined. From a different view, virtual reality experiences can be used to change people's mindset about viewing a certain area of a company's operations. Essentially, rethinking creative thinking helps company personnel put on a "new set of eyeglasses" so that they can see the world in a new and a different way from the past.

To assist company personnel in terms of what needs to be done can be facilitated by getting them to employ creative techniques online. According to a study by PricewaterhouseCoopers, almost half (45%) of creative ideas — whether they be breakthrough products or services, new ones to replace old ones, or ways to cut costs —- come from company employees. By contrast, customers, suppliers, and competitors contribute approximately the other half. A growing number of companies, like Georgia-Pacific, W. R. Grace, and ChevronTexaco are taking advantage of the Internet plus specially designed software to run brainstorming sessions so that all of their employees located throughout the world can "meet online" and determine new market opportunities for goods and services as well as determine solutions to specific problems. In effect, creative thinking technology allows employees to see and build upon one another's ideas so that each person's seed of a notion or idea can grow into a practical plan for the given opportunity or solution to a problem under study. In a global economy where complexity is the order of the day, this online approach to creativity allows companies to move to the next level for determining what really needs to be done organizationally over time (Fisher, 2004).

Focus on the "Big Picture" by Asking the Right Questions for What Needs to be Done Over Time

As discussed in the prior chapter, an optimal knowledge management/wisdom management system allows a company's decision makers the ability to judge soundly so that optimal decisions can be made for growing an organization over time. It assists decision makers in accumulating wisdom needed to understand the opportunity or the problem, thereby cutting to the heart of the relationship by connecting points of wisdom. Its focus for decision makers is getting a good handle on the "big picture" by asking the right questions for what needs to be done today and tomorrow. In the past, decision makers tended to find answers to questions that had some or very little impact on the organization's total operations. Today, there is need to center on the collective wisdom of decision makers so that the real questions affecting operations are asked and answered to the satisfaction of the organization's decision makers. Needless to say, this approach requires new insights into the organization's operations by decision makers to seek out these big picture questions. In turn, related questions can be asked and answered that allow the right hand to know what the left hand is doing.

Basically, the purpose of questions is to assist both the questioner and those being asked in finding answers to the areas under study. From a positive perspective, questions are means to assist — not to demean — and will help everyone involved to find answers. Just the act of asking makes other people feel that one cares and wants to listen to what they have to say. It creates a positive feeling in people and will allow everyone to work together to find the best answers. When someone goes off track, asking the person a direct question is a good way to help him or her stay on his or her original path. Thus, questions can be used to help others find answers for themselves rather than answering for them. This creates self-esteem and empowers them to do more for themselves. It also frees the person to do his or her own job rather than taking the time to help others do theirs (Goldsmith, 2002).

As an example of getting the big picture about what needs to be done, reference can be made to Microsoft where Bill Gates has been asking the right questions of his smartest people to grow the corporation. More specifically, he hires the smartest people possible and then gives them the resources to transfers their ideas into desired results. As Microsoft has grown this culture of innovation ensures that Microsoft has the processes and infrastructure to turn ideas into products on a very large scale. By never forgetting the big picture, Microsoft has been able to achieve a good balance among fundamental research, incremental development, and innovative products. In addition, since the company is so large, it has the ability to make large investments in innovative ideas that can deliver important new capabilities desired by its customers today and tomorrow (Gates, 2004).

Ways of Boosting a Decision Maker's Brain Power

To assist decision makers in asking the right questions that focus on the big picture, there is need to boost their brainpower. As a starting point, decision makers should take the time to really think about the answers to key questions about an opportunity, or problem, or some situation. That is, what seems to be the key idea here? Does this resemble or parallel anything learned or experienced previously? When everything is put together, what do decision makers see to be the important aspects? Another approach to boosting a decision maker's brainpower is to reverse the questions from a positive viewpoint to a negative one so that all aspects of the question are viewed from a big picture perspective. Another approach is to view the questioning in terms of pictures, graphs, and the like. It is particularly helpful to view the question in terms of its highs and lows in the past and currently, which may indicate areas for improvement. Going one step further, future projections for the question being studied may indicate forthcoming opportunities and problems that need to be recognized today. Yet still another approach to boosting a decision maker's brainpower centers on viewing individuals as groups. Many times, people as individuals do not follow any pattern, but people as groups can be analyzed and their behavior predicted to a large degree. Lastly, a number of creativity techniques can be used to assist decision makers in boosting their brainpower to solve important problems facing the organization and explore opportunities that will allow it to grow.

Effective Techniques to Undertake Creative Thinking

Today, there are a number of effective creativity techniques to assist decision makers in boosting their brainpower. As a starting point, the challenge for decision makers is to think up ideas that are novel and innovative and, at the same time, are pragmatic. This type of creativity requires being open to new ideas while avoiding the problem of coming up with the same answers. If company personnel continually come up with the same solutions, it could be that those are the only answers or that they are only looking at the situation from one perspective. Companies are buying into the notion that rather than being the province of artists and inventors, creativity is a learnable skill that can be enhanced through practice and training. Proven creativity techniques, creative computer software, and electronic collaborative creativity are set forth below.

Typical Proven Creativity Techniques

Typical proven creativity techniques include: (1) brainstorming, (2) synectics, (3) accurate problem definition, and (4) rapid prototyping. These techniques are set forth below along with examples. Basically, these techniques can assist managers and their support staffs in getting around the need to be right all of the time, which can be a significant barrier to developing new ideas. It is better to have some new ideas, which may prove to be wrong versus having no ideas at all.

One of the best-known creativity techniques is brainstorming. It was developed by Alex F. Osborn (cofounder of BBD&O) to help solve advertising problems. Its focus is on improving problem analysis by providing more possible solutions and unusual approaches to the problem under study. Osborn suggests the following rules for the utilization of brainstorming: judgment is withheld such that ideas may be criticized and evaluated later; wild ideas are encouraged since ideas are easier to modify than to originate; numerous ideas increase the possibility of obtaining an excellent idea; and the participants are encouraged to utilize the ideas of others to develop additional ideas. Additional brainstorming procedures include: the sessions should be recorded because some ideas may be missed during a meeting; the problem must be manageable, even if it requires breaking large problems into smaller parts; and samples should be available if products are being discussed. Once the brainstorming session has ended, the group must set up the criteria for evaluation. Next, all the ideas are evaluated based upon the criteria and the best two or three possible solutions are chosen.

In the final analysis, there may be need to modify any idea in order to make it more closely meet the desired criteria. A seemingly wild notion can be transformed into a workable solution.

Synectics, which is based on the assumption that creativity can be described and taught, is useful to improve the quality of creative output from those assigned to a synectics team. Essentially, the synectic process centers on making the strange familiar and making the familiar strange. The first item requires that the problem be understood and that the ramifications be considered. The mind tends to emphasize one's own experiences and to force strange ideas into an acceptable pattern. Basically, it is necessary to re-orient these strange ideas into familiar ones. The second item, making the familiar strange, involves distorting, inverting, and transposing the problem in an attempt to view the problem from an unfamiliar perspective. Reference can be made to a book by W. Gordon for more details on synectics (Gordon, 1961).

Because many problem-solving failures occur when efforts are directed at solving the wrong problem or only parts of it, there is need to identify accurately what is really going on. To assist in accurate problem definition, that is, defining the real problem, a cause-and-effect diagram is recommended. Because problems cause other problems, a whole complex of symptoms and problems emerges in need of a solution. Hence, it is helpful to utilize a problem diagramming procedure that isolates root causes. Initially, list all the problems, symptoms, and related problems by numbering each one. Next, write the numbers at random on a piece of paper and draw a circle around each. Then draw arrows to show what causes what. If problem 1 causes problem 2, for example, draw an arrow from circle 1 to circle 2. When considering each circled number, focus on which of the other problems causes or helps cause this one. After all the arrows have been drawn, the root problems become clear and can be represented by the circles with arrows leading only away from them. Finally, the real root problem is represented by the farthest one to the left.

Another creativity technique is rapid prototyping states that it is much easier to discuss a model of something, no matter how primitive, than to talk about a number of ideas. If a picture is worth a thousand words, a prototype can be worth several thousand. Basically, rapid prototyping consists of three Rs – Rough, Rapid, and Right. The first two Rs are fairly self-explanatory - make the model rough and make it rapidly. In the early stages, perfecting a model is a waste of time. The final R, Right, does not mean the model has to work. The focus is on building a number of small models that focus on specific problems. Hence, the creator is not trying to build a complete model of the product but rather the individual is focusing on a small section of it (Brown, 1999).

Creative Computer Software Packages

Today, there are a number of creative computer software packages, two of which are given below. IdeaFisher 4.0 (from Fisher Idea Systems) is a kind of brainstorming thesaurus with two databases that work together. The "QBank" is a collection of about 6,000 questions that nudges the user to define the job at hand. The user might be quizzed, for example, about the engineering of a proposed product, its timeliness, and the history of similar products. The "IdeaBank" contains more than 700,000 cross-referenced words and phrases organized into 28 "major categories" and 387 "topical categories." Typing "blue" triggers a number of associations that offer fodder for everything from advertising slogans to new product names, that is, midnight blue, blue whale, and so forth. As an example, a marketer is trying to name a new laundry detergent and wants to convey the message that the product is efficient and environmentally sound. The individual would enter such terms as "environment", "clean", and "detergent", and the IdeaBank would give back suggestions that included Breathe, Back to Nature, and Purify.

Idea Generator Plus (from Experience In Software) is another creative computer software package. This software makes the user examine the problem from unexpected angles by having the person respond to such questions as: "What similar situations have you been in?" "What is the opposite of what you want to achieve?" "Can you think of metaphors that apply?" "What does the pessimist in you think?" "Who are the people affected by your decision and what solutions might they offer?" A consulting company, for example, used this package to devise an adult board game in which the players pretend to be inventors.

Within an optimal KM/WM system operating mode, creative computer software, can be an important force to assist decision makers in devising new ways to explore opportunities for changing times. Creative computer software packages can add a new dimension by enlarging managers' capabilities to develop new opportunities for their organizations. Reference can be made to a prior publication on creative computer software by one of the authors (Thierauf, 1993).

Electronic Collaboration to Aid the Creativity Process

Creating value across a corporation requires contributions from every employee. In addition, employees who are satisfied and focused are more productive when creating values for its customers. From the standpoint of electronic collaboration to aid in the creativity process, Smart Technologies Inc.'s Smart Ideas 2.0 tool marries Internet conferencing with graphical diagramming tools to create highly flexible collaboration software. Smart Ideas is one of a number of new graphical groupware applications, including MindJet LLC's MindManager 3.5 (as noted below) that allows collaboration using Java-based clients linked via TCP/IP. Smart Ideas is one of the better products for larger groups of users because it taps a server to centrally store collaborative data. Available as either a stand-alone Java application or a plug-in to a Web browser, Smart Ideas has a client-user interface that is easy to use. It produces diagrams known as concept maps that divide complex ideas into components and display the relationships between ideas. Each idea can be placed in its own box and related concepts are linked with lines. Because Smart Ideas' server stores maps and related documents, participants can work on a project at their convenience or use the server for collaboration. Throughout the electronic collaboration, changes can be viewed by other participants made in the map.

For more controlled collaboration, Mindjet LLC's MindManager 3.5 has a good range of information sharing options. MindManager is capable of producing concept maps to display ideas and the relationships between concepts. It produces lavish illustrated maps and has more controlled conferencing and information publishing options than its rivals. Since it lacks a server, it is more appropriate for small groups. MindManager 3.5 produces highly detailed concept maps with images and symbols. Its wide variety of information publishing options let users share these maps with other MindManager users, publish maps on Web servers as HTML images, and send them via e-mail as bit maps.

Creative Thinking Example to Assist in the Development of an Optimal KM/WM System

Utilizing the above techniques plus others, creative thinking is widely used by innovative decision makers for developing optimal KM/WM systems as well as growing them. As leaders with high creativity and imagination, innovative decision makers find unique solutions to opportunities and problems. They throw out conventional thinking and turn ideas into practical solutions. Although they are mavericks, they have new ideas that are realistic and can work. Truly effective innovative decision makers follow through on the transformations they help to launch. As agents of change, they see conflict as potentially healthy and perhaps useful in developing the leverage needed for changing times. The bottom line is that innovative decision makers often turn to creative thinking and their related techniques for pruning their organizations. Innovative and creative organizations are developing and implementing the basics of optimal KM/WM systems, one of which is set forth next.

In the late 1990s, FedEx Corporation customers were clamoring for a quick and easy way to navigate the fast-paced world of international trade. In a few words, its customers were asking FedEx to revise its basic operating business model. So, Robert B. Carter, Executive Vice President and Chief Information Officer at FedEx, assembled a team and set about developing FedEx Global Trade Manager, a free, Web-based guide to international shipping for small and midsize businesses. It was designed around its customers' needs for changing times. FedEx designed its own Web-based system and integrated it into an Oracle Corporation database. The application helps shippers understand global trade regulations and prepare the appropriate import or export forms based on the commodity being shipped and the countries of origin and destination. It also alerts users to restrictions on shipping certain commodities, lets them know if a country is under embargo, and provides information on special licensing requirements (Rosencrance, 2002).

Although this system has worked well for the times, the FedEx Global Trade Manager is extending its capabilities by incorporating some of the concepts from optimal KM/WM systems. More specifically, this means going outside the organization and listening very intently to its customers. FedEx is redesigning its operations such that there is an open flow of new ideas from its customers. What may work well in one part of the world may not work well in another part of the world. In this manner, FedEx's decision makers can judge soundly on the effectiveness of its operations around the world. Selected changes can be made where appropriate to meet the changing demands of the world market.

Problem Finding Useful in Turning Problems into Opportunities for Profit

To assist innovative decision makers in getting an overview of a company as well as its detailed operations for developing and growing optimal KM/WM systems, problem finding and its related techniques are useful in getting a handle on problems and opportunities arising out of them. Typically, company decision makers need to stand back and take a very broad-based holistic view by distinguishing between a reactive and a proactive approach to resolving problems and exploring opportunities. Today, too many managers spend more time putting out fires and less time in helping their business grow. Typically, a company has taken a reactive approach to its problems, that is, as they arise, they try to apply appropriate management techniques in order to resolve current problems. In reality, several more problems tend to crop up that require the attention of management. This never-ending scenario of fighting problem fires never seems to get under control.

On the other hand, a better approach is to take a proactive, preventive approach that keeps managers on top of problems confronting them. Generally, these problems were always there but management either ignored them hoping that they would go away or assigned them to others who do not have the time or clout to get them resolved. Hence, the preferred approach today is to have managers get involved in problem finding versus just problem solving.

The problem-finding process not only includes anticipating future company problems and bringing them back to the present time for solution, but also looking for future opportunities that are related to these future problems. In marketing, for example, problem finding includes knowing what customers want before they know themselves versus just meeting competition. In finance, this means that risks can be managed with foresight versus damage being controlled through hindsight. In manufacturing, problem finding can be used to maximize supply chain relationships by developing new opportunities for all parties. This proactive approach to a company's problems and opportunities is an important prerequisite for improving a company's competitiveness in today's fast changing times. Also, the problem-finding approach can be tied-in with a company's critical success factors, that is, upcoming problems and opportunities that impact a company can be related to those factors that are critical to a company's survival.

To assist company employees in problem finding, it is essential that senior-level employees' deepest knowledge and intelligence about an organization's operations be transferred slowly, patiently, and systematically over time to their juniors. Essentially, employees do not learn from a series of computerized presentations, Web sites of best practices, online training programs, project reports, or lectures. But rather, company employees learn to draw insights from their own experiences by working with senior level personnel. Employee deep smarts are based on know how and not just on facts alone. They comprise a holistic view of a company's operations as well as its functional areas. The bottom line is that an organization needs to allocate the required time to such efforts such that company employees can learn to draw proactive and realistic conclusions from their experiences over time, that is, the ability to judge soundly. In this manner, problem finding can be central to their thinking for turning problems into opportunities for profit (Leonard & Swop, 2004).

The Problem Finding Process Found in Optimal KM/WM Systems

Underlying the problem-finding process for optimal decision making is "logical analytical thinking" that centers on identifying and solving future problems. In actuality, future problems are actually future opportunities in disguise. In the solution of future problems and/or opportunities, it is helpful to employ one or more of the creativity techniques found previously in the chapter. Overall, a manager who has identified future problems has also identified opportunities. From this view, the problem-finding process can be broken into two parts — a problem-centered approach and an opportunity-centered approach for optimal decisions today and tomorrow.

In the problem-centered approach, logical-analytical thinking centers on examining the environment with the idea of looking into the future and exploring problems that will have an impact on the organization now or at some time in the future. Typically, the process is centered on projecting into the future, determining important problems (i.e., problem finding), and bringing them back to the present time to examine their cause-and-effect relationships. Today, this includes the impact of problems on the organization's business model and the various optimization techniques used by the organization on a day-to-day basis. Logical-analytical thinking is also needed in the opportunity-centered approach where the main focus is on identifying opportunities for the organization to pursue that generally comes from problems uncovered. As such, managers do need to change an organizational liability into an asset, that is, identify problems that can result in important opportunities for the organization. Since the opportunity-centered approach need not always be related to future problems, it can assist top management and/or the corporate planning staff on identifying current and future opportunities. Also, this approach is extremely helpful in assisting management in determining what the business model should be for the organization and what optimization techniques can be helpful to the organization. New opportunities may signal the need to change or possibly scrap the current organization business model and related optimization technique to ensure optimal decision making.

First Approach is Problem-Centered

The first approach which is problem-centered is taken from the authors' previous publications with some modifications, if appropriate, to include the generation of new ideas using creative computer software (Thierauf & Hoctor, 2003). As shown in Table 1, this problem-centered approach consists of four steps plus the solution and implementation steps from the problem-solving process. Typically, these four steps precede the actual solution to the future problems under study. Overall problems must be identified before they can be solved on an optimal basis.

Table 1. A comparison of steps in the problem-finding process — the problem-centered approach and the opportunity-centered approach (Note: There is feedback from the last step of both approaches to the first step)

Basic Phases	Problem-Centered Approach	Opportunity-Centered Approach
Search	Step 1 Generation – probe for potential problems that might exist in the future using a brainstorming approach.	Step 1 Exploration – examine the environment for opportunities that come from problems uncovered using a brainstorming approach. If appropriate, use creative computer software to explore new ideas to exploit these opportunities.
	Step 2 Evaluation – review problems uncovered for managerial concern, backed up by a cost-benefit analysis. If appropriate, use creative computer software to generate new ways to evaluate future problems.	
Identification	Step 3 Validation – select actual problems for managerial concern. If appropriate, relate to the company's critical success factors.	Step 2 Selection – determine that one or more opportunities should be explored by management. If appropriate, relate to the company's critical success factors.
	Step 4 Establish Boundaries – define each potential problem within its boundaries so as to cover the whole area that the problem encompasses.	Step 3 Examine Boundaries – survey the environment for each opportunity and determine the proper boundaries.
Solution	Step 5 Development of Appropriate Solution to Problem – solve for an optimum or, at least, a good solution	Step 4 Development of Best Solution for the Opportunity – solve for an optimum or, at least, a good solution.
Implementation	Step 6 Implementation of Solution – includes verification and establishment of proper controls.	Step 5 Implementation of Solution – includes verification and establishment of proper controls.

Generation

Since problem search centers on a probe of potential problems that might have an impact on the company initially, the analysis is "forward-looking". It is a search for future problems. Once these problems are identified, the analysis becomes "backward-looking" since there is a need to evaluate the cause-and-effect relationships of each problem and its possible effects on the organization currently. Accent is placed on each problem, from the short range to the long range. In addition, it may also be necessary to look at each problem in different economic climates (good, average, and bad conditions).

The best approach to generate important problems is to use brainstorming. Typically, top managers and their staff, along with members from the corporate planning staff, meet periodically to brainstorm future organizational problems. In a typical session, all-important problems uncovered are recorded; then analysis is performed to explore the important aspects of each problem. These steps are performed in a back-and-forth fashion, that is, the original question concerning the problem as given and the subsequent spontaneous ideas are all written down. When participants' minds have cleared, they concentrate on reformulations produced from the collected material, and a choice of one or more is made before continuing with questions in sequence concerning the problem as understood. (As noted, the initially forward-looking analysis becomes backward looking.) This can be repeated until all aspects of the problem have been covered.

Evaluation

The second step centers on examining problems in terms of their being worthy of managerial concern. For problems that are found in the future — the next two to five or possibly ten years — the question can be asked, "Which problem or problems should be undertaken for solution?" To answer this question, there is need to evaluate the impact the solution to a problem has on the organization, especially in terms of net profit and return on investment. In other cases, consideration might be given to other important areas of an organization, such as customer sales and service. Similarly, it may be necessary to relate the problem or problems back to the organization's critical success factors, which, in turn, affect its key performance indicators and financial ratios. Many times, it is helpful to perform a cost-benefit analysis to determine the impact of the solution on the financial aspects of the organization today and tomorrow. This task, for example, can be relegated to managers at the appropriate levels and their staffs to determine which problems are of valid concern for managerial action. If deemed appropriate, creative computer software can be used to generate new ways and ideas that are related to help in the evaluation of future problems. The problems generated, then, are evaluated in terms of benefits versus costs.

Validation

Actual problems are selected as being worthy of managerial concern for today and tomorrow. The validation for solving these problems is generally backed up by a cost-benefit analysis and related to the company's critical success factors if appropriate. If such an analysis is not available or much too difficult or very costly to develop, it may be necessary to use alternative means, such as the consensus of the majority of this problem-finding group to substantiate this selection as an important problem to be solved. For the most part, the staffs have prioritized the important problems to be solved. It is up to the problem-finding group to pass judgment on them.

Establish Boundaries

At this time, it is necessary to describe (define) each problem within its boundaries. This ensures that areas that the problem might touch or come into contact with will be included in the problem-finding process. The net result is that there is need for some fine-tuning such that the appropriate boundaries of the problem will be considered in its solution. Typically, to establish realistic boundaries, the problem-finding group must have a rich knowledge of the future (good, average, or poor economic conditions), a clear description of performance that a solution must fulfill, and a clear idea of what to expect from solving the problem. These areas must be as clear and as accurate as possible. Otherwise, if the problem is badly defined, the solution is generally of no value to the organization.

Development of Appropriate Solution to Problem

Solution to the problem-centered approach can take one of two directions. One is the development of alternative solutions and selection of best solution for well-structured problems. The other is selecting a good solution from a set of feasible ones and evaluating courses of action for solving unstructured problems. In either case, the solution centers on solving the problem before it actually occurs where the accent is on practicing "management by perception." This is opposed to using "management by exception" as found in the past.

Implementation of Solution

Implementation includes verification of the optimum solution and establishment of proper controls for well-structured or unstructured problems. It should be noted that there is need for feedback in the problem-centered approach from the last step back to the first one to ensure that what was intended was actually undertaken.

Second Approach is Opportunity-Centered

This second approach which is opportunity centered is also taken from a prior publication by the authors (Thierauf & Hoctor, 2003). As illustrated in Table 1, the opportunity-centered approach consists of three steps plus the solution and implementation steps from the problem-solving process. Like the problem-centered approach presented above, these three steps precede the actual solution to the future opportunity under study.

Exploration

The internal and external environments are examined for opportunities that can come from problems uncovered. As in the problem-centered approach, brainstorming is generally used by managers and their staffs. The focus is directed away from the short range to the medium and long range where every effort is used to determine what opportunities are presented by the problems discovered in the future. As in the problem-centered approach, the analysis is initially forward looking in that there is a search for future problems. Once the problems have been identified, they are examined from the standpoint of identifying opportunities for improving the company's operations (from the standpoint of sales, profits, or whatever). As such, the concept of opportunities has a "positive" connotation while the concept of problems has a "negative" meaning. If deemed appropriate for the situation, creative computer software can be employed to assist managers and their staffs in developing new ideas that exploit appropriate opportunities for a company to pursue.

Selection

The appropriate opportunities are determined in terms of which ones should be explored by managers and their staffs. The selection process should focus on opportunities that can relate to a company's critical success factors, the factors that critical to its success. Typically, these factors include selling price, sales promotion, customer service, product mix, inventory turnover, cost control, and quality dealers. In turn, the interrelationships of the critical success factors and the company's goals and objectives are discussed for further clarification. The purpose of this discussion determines what opportunities should be pursued by the company, thereby identifying them in a clear and meaningful way. Moreover, it takes into consideration all the important facts that bear on important company opportunities. Where deemed necessary, a cost-benefit analysis can be used to determine what opportunities are more important than others in terms of potential profits and return on investment.

Examine Boundaries

The environment should be surveyed for opportunities identified before pursuing an opportunity solution. Due to the nature of some opportunities, the boundaries may be quite wide, that is, they may extend beyond the company and may be related to emerging and established organizations and industries. Typically, greater opportunities are found when boundaries are extended. The bottom line is that top management and the corporate planning staff need to examine the boundaries surrounding the opportunities from a narrow to a very wide perspective.

Development of Best Solution for the Opportunity

Like with the problem-centered approach, the solution to the opportunity-centered approach can take one of two directions: determination of the optimum solution or the selection of a good solution from a set of feasible ones for the opportunity.

Implementation of Solution

Implementation includes verification of the optimum opportunity and the establishment of proper controls for making the necessary adjustments to changing times for the opportunity selected. Feedback in the opportunity-centered approach goes from the last step back to the first one.

Utilization of Problem Finding to Expand Wisdom of Decision Makers

The problem-finding process — as discussed in this chapter — centered initially on getting a grasp on an organization's future problems and related opportunities. In addition, the problem-finding process utilizes information, knowledge, and intelligence that is useful in implementing new ideas to solve future problems as well as undertake future opportunities for an organization. It should be noted that a thorough understanding of information, knowledge, and intelligence gives decision makers the ability to learn, interpret, and deal with new and different situations for applying optimization techniques to a company's resources for optimal decision making. This larger point of view displaces the prior belief that more information, knowledge, and intelligence will solve a company's problems. As such, information, knowledge, and intelligence along with optimization techniques underlie new concepts and ideas and their detailed analysis leads to a better understanding of them. This understanding allows for making sound judgments about the organization and its related operations for what needs to be done.

Building upon these basics of problem finding, decision makers have come to realize that incremental, process-specific change programs are no longer enough. On the other hand, they know that undertaking an organization-wide change program, such as a product lifecycle management approach, is an enormous and complex challenge. In both cases, there is need to align the organization's business model to these changes. A large-scale change program needs the guiding force of wisdom — the ability to judge soundly that is gained from the decision maker's vast experience over time. Hence, wisdom is needed to translate that understanding into specific changes in organization structures, culture, business operations and processes, and information systems to implement and maintain an optimized organization. Otherwise, the effort is like attempting to build a skyscraper without a blueprint. In terms of understanding the business, many organizations, locked into their traditional ways, have difficulty in articulating exactly what they do (i.e., the work of the business) or why they do it (i.e., the purpose of the work). To assist in gaining wisdom, modeling of the whats and the whys of the organization's work can be undertaken by systems designers. It is then possible to build a stable blueprint that can be used to align the entire business. This approach is based on the premise that every business has one fundamental purpose and that all work performed in the organization should have a demonstrable link to that purpose. Once the business purpose is defined, the modeling of the work is approached first at the various levels of the business by asking what work is done and why it is done, independently of existing people, processes, volumes, technology, and other business specifics. Only work that supports the fundamental business purpose is included in the model within an optimal KM/WM environment. Throughout the process, decision makers are assisted by a problem-finding approach to expand their wisdom for a holistic view of a company's operations for optimal decisions.

Because problem finding and the employment of wisdom by decision makers focus on future opportunities that can be related to work, several companies have adopted a new approach to product research and development. More specifically, these organizations have abandoned their efforts to understand exactly what products their customers want. Instead, they equip their own customers with tool kits to design and develop their own products. Not only must a company develop the right tool kit, but it must also revamp its business model and management mind-set. When companies relinquish one of its fundamental tasks such as designing new products for its customers, the two parties must redefine their relationship. For example, with custom computer chips, companies traditionally captured value by both designing and manufacturing innovative products. With customers taking over more of the design, companies must now focus more on providing the best custom manufacturing. Essentially, the location where value is created and is captured changes, and business models for both must be reconfigured. Thus, the development of new opportunities can lead both buyers and sellers into new relationships that were not envisioned before. As such, combining problem-finding capability with the wisdom of decision makers makes for a win-win situation not envisioned in the past (Thomke & Von Hippel, 2002).

Example of Problem Finding Within an Optimal KM/WM System

To illustrate problem finding within an optimal KM/WM system, the following example will be used for both the problem-centered approach and the opportunity-centered approach. Consider a pet foods company that is currently offering a premium line of cat and dog foods, which are sold exclusively through pet food stores. To illustrate the problem-centered approach, recently the president and the marketing vice president have become concerned about the entry of new premium brands by the major suppliers of pet foods. Upon brainstorming the overall problem with the company's chief corporate planner, a number of problems were generated (first step) by these high-level decision makers. One of these is the effect of reduced volume since other pet food suppliers are offering the same type product at the same or lower prices. An equally important problem is the need to change selling strategies aimed at a more affluent buyer segment of the market. Needless to say, other problems that are tied to these two problems can be identified. In turn, ideas related to overcoming these problems are evaluated using a creative computer software package. In addition, a cost-benefit analysis that is related to a company's objectives and stated financial goals serve as a financial evaluation basis (second step).

At this point, the president, the marketing vice president, and the corporate planner met to validate the overall problem of increasing competition and the related problems of lower sales volume that is tied-in with the potential lower profits (third step). After examining their impact on customers, production, and other concerns, priorities that are related to the company's critical success factors are set forth for the specialized problems areas. Next, the scope of the boundaries for each problem is addressed (fourth step). As examples, the marketing efforts are defined in terms of advertising medias and the markets to address. Changes in production and inventory levels are outlined for the manufacturing facility. Other boundaries, like personnel, budgeting, and time frames are set forth for each problem.

Possible solutions for each of the problems are identified (fifth step). Solutions for the marketing department include a more aggressive advertising program aimed at more affluent buyers of pet foods or developing newer pet products to reflect current trends. Solutions for the manufacturing department include having existing plants expanded to newer manufacturing methods to reduce costs or contracting the work to outside suppliers within or outside the country. Additional solutions can be related to budgeting constraints and personnel factors. After analysis of potential solutions, these three top-level decision makers need to decide what is the optimum solution for each problem after considering all of the factors. The optimum solution for each problem needs to be implemented (sixth step). In addition, there is need to place proper controls over the implemented solution(s). Finally, there must be a process for feedback back to the first step of the problem-centered approach.

Moving on to the opportunity-centered approach for this pet foods company, the president, the marketing vice president, and the corporate planner need to examine the environment for opportunities that come from problems uncovered using a brainstorming approach. More specifically, they need to identify future opportunities that may exist in terms of new pet products that focus on dogs, cats, or both. That is, it is not enough to tell owners about some new aspect of pet foods, but also allow their owners to have a new experience that was not possible before. This could take the direction of newer products that demonstrate to pet owners that their pets have more luxurious fur or their pets have more energy than with past products. In effect, top-level decision makers must help their customers view their pets in a new light, that is, in a way not offered by other pet food suppliers (first step). Utilizing the same creative computer package as set forth previously, important ideas, like the above, can be used to exploit important new market opportunities.

By looking at the tradeoffs of each opportunity, one or more is selected for further review. Using a cost-benefit analysis, a newer product that incorporates the above attributes is determined to be the best alternative to beat or, at least, meet competition. At this point, it is helpful to relate this new pet food to the company's critical success factors (second step). The boundaries of this new product are examined (third step), that is, does this pet product really interest a wide range of pet owners or only a very small segment, does the introduction of such a product improve the company's profitability, or does the product really meet the actual needs of pet owners? These boundaries will lead to the best solution (fourth step). The implementation of the solution includes verification and establishment of proper controls (fifth step). Lastly, there is need for feedback to the first step of the opportunity-centered approach.

Summary

In light of the ever-changing economic times in which business organizations find themselves, employing creative thinking and problem finding for optimal decision making have become even more important today, especially when building optimal KM/WM systems that chart new directions for a company. Initially, the chapter looked at creativity as part of a company's corporate philosophy plus the need to rethink creative thinking in terms of what needs to be done. The focus on the big picture by asking the right questions for what needs to be done was examined, followed by effective techniques of all types to undertake creative thinking. Next, the essentials of problem finding were examined and tied-in with optimal KM/WM systems for changing times. Problem finding not only looks at future problems and brings them back to the present time for solution, but also examines future opportunities that have the potential to alter the direction of an organization in a minor to a major way. Overall, the main thrust of creative thinking and problem finding as found in optimal KM/WM systems is to align the organization better to changing times for optimization of an organization's resources.

References

Brown, E. (1999). A day at Innovation U. *Fortune*, 163-165.

Fisher, A. (2004, November 29). Get employees to brainstorm online. *Fortune*, 72.

Gates, B. (2004, December). How to exploit innovation. *Business 2.0*, 105.

Goldsmith, B. (2002, June). The awesome power of asking the right questions. *Office Solutions*, 52-53.

Gordon, W. (1961). *Synectics: The development of creative capacity*. New York: Harper & Row.

Leonard, D., & Swop, W. (2004, September). Deep smarts. *Harvard Business Review*, 88-97.

Rosencrance, L. (2002, March 11). Web guide remits rules of shipping. *Computerworld Premier 100 Best in Class*, 15.

Thierauf, R. J. (1993). *Creative computer software for strategic thinking and decision making: A guide for senior management and MIS professionals.* Westport, CT: Quorum Books.

Thierauf, R. J., & Hoctor, J. J. (2003). *Smart business systems for the optimized organization.* Westport, CT: Praeger.

Thomke, S., & Von Hipple, E. (2002, April). Customers as innovators: A new way to create value. *Harvard Business Review*, 74-81.

Chapter III

Computer Storage and Networking Found in Optimal KM/WM Systems

Issues

- To explore future computer technology that will assist in the development of optimal KM/WM systems
- To look at business process management for connecting "points of wisdom" and information lifecycle management to oversee computer storage growth
- To examine computer storage that is conducive to developing and maintaining an effective optimal KM/WM system
- To explore the world of wired and wireless computer networking that is tied-in with the development and growth of this newer operating mode

Introduction

Initially, future computer technology that will assist in developing optimal KM/WM systems is explored. Also, business process management and information lifecycle management are examined. Effective *computer storage* follows next that centers on storage of *aged data* along with the need for a data federation approach for *real-time data*. The types of local and corporate wide databases are examined. In the second part of the chapter, the focus is on *computer networking* that includes wired and wireless technologies. There are a number of topical areas covered, including the Internet and the World Wide Web along with e-commerce. Typically, networking operations must be managed with greater levels of reliability and security than in the past. Overall, a discussion of computer storage and networking provides a background for developing and growing optimal KM/WM systems over time.

Future Computer Technology that will Assist in Developing Optimal KM/WM Systems

To better understand the computing environment — computer storage and networking in this chapter and computer software in the next chapter — found in an optimal KM/WM system-operating mode, a future look at their related computer technology would be beneficial. For one, *ubiquitous computing* which is the ability to access digital content, both personal and business related, through a variety of digital information devices will supplant the personal computer. It will evolve such that it ultimately displaces the personal computer on the desktop. An integral part of ubiquitous computing are three components: (1) a multitude of information devices, (2) connectivity between personal-area networks, wireless LANs and WANs, and (3) Web-based services that provide the software infrastructure. Additionally, the Internet will play a major role in the widespread use of ubiquitous computing (Gartenberg, 2001).

Related to ubiquitous computing is *utility computing* which combines hardware, software, and services into a total utility service that provides computing resources at a very low cost. The impetus for this approach to computing is based on the fact that companies, on the average, underutilize server resources and systems rarely exceed 20% of their workload capacity. Essentially, utility computing allows a company to plug-and-play resources that scale up or down as a business needs them to and that cost will be based on their usage. This pay-as-you-go utility model allows a company to run certain of its information department operations safely elsewhere, thereby reducing processing costs and freeing up company personnel for other company operations. For example, IBM and HP think buyers will want to get rid of as much of their computer department operations as possible. In turn, they can deliver networking, servers, applications, and storage that companies plug into just like electricity (Rist, 2005).

Another new source of future computer technology is *grid computing*, also known as the Grid. The Grid gives users access to unprecedented computing power and data no matter where they are located. Ultimately, the Grid will be like having a supercomputer at one's fingertips. The Grid is more than just a physical infrastructure, that is, computers, storage devices, and the networks, that connect them in a way that roads, pipelines, and power lines connect cities. Grid computing also requires highly sophisticated software programs that enable a user at point A to draw on the computational resources located at points B, C, D, and so forth. While the Internet lets computers talk together, grid computing lets computers work together. The Grid will find a place in complex computing activities, such as investment risk analysis, oil exploration, and complex mechanical design. Grid computing will be helpful for scientific research of all types (Ricodela, 2005).

Future computer technology will also include *wearable computers*. A wearable computer is a computer that can be worn by a human being and that has full PC functions, including keyboard, flat screen, power source, and so forth. Universities, like Carnegie Mellon, Georgia Tech, and MIT, have on-going programs devoted to wearable computers. Their interest has been augmented by organizations such as Boeing, DARPA, IEEE, and a collaborative effort between Creapole Ecole de Creation and MIT on the impending marriage of wearable computers and fashion. The military has also shown considerable interest. Computer vendors are currently offering a wearable computer. For example, Xybernaut's Mobile Assistant is a wearable computer that makes use of a head-mounted display, a Pentium MMX CPU, and voice recognition and activation. It is able to access files, send e-mail, and connect with the Internet. Although the technology of a wearable computer is well along, its most difficult part will be scale of function, weight, sustained power, and cost (Malone, 2001).

In addition, future computer technology can be related to artificial intelligence (AI). Faster and cheaper computer processing power, memory, and storage, and the rise of statistical techniques for analyzing speech, handwriting, and the structure of written texts are helping to generate interest in new developments. Researchers are building AI-inspired users interfaces, systems that can perform calculations or suggest passages of text in anticipation of what users will need. Much of the research employs Bayesian statistics — a branch of mathematics that factors in common beliefs and discount surprising results in the face of contrary historical knowledge. Several industries can benefit from AI research results. The emerging field of wireless sensor networks which collect vast amounts of data about industrial operations, the ecosystem, and the conditions in a building or home. The Pentagon finds AI research helpful to lay the groundwork for intelligent vehicles and robots (Ricodela, 2005).

Looking even further into future corporate technology, futurists are predicting that computers will be able to read information stored in the human brain. One futurist, John Norseen, foresees the computer's capability to convert a person's thoughts into computer commands by deciphering the brain's own neurotransmitters, for example, to fight off disease, enhance learning, or alter the mind's visual images, thereby creating what Norseen has dubbed "synthetic reality" (Pasternak, 2000). In addition, Ray Kurzwell — the author of *The Age of Spiritual Machine* and a prominent inventor and business leader in the field of artificial intelligence — believes by the year 2030 that a $1,000 personal computer will achieve the full capacity of the human brain. This prediction is based on the ongoing exponential growth of computer technology. Currently, computers remember trillions of facts faultlessly and are also much quicker that the fastest intellect. They are able to search a database with billions of records in a fraction of a second. Kurzwell predicts that 25 years into the future, computers will have this capability plus others. This includes consciousness and the ability to have emotional and even spiritual experiences that have not been seen in any form in the past (Goldsborough, 2000).

All of these future capabilities of computer technology, needless to say, will have a profound impact on the development and growth of optimal KM/WM systems. As computer technology evolves to better meet the demand of a company's decision makers, it is expected that decision makers will accumulate enough wisdom to explore the most promising opportunities thoroughly and see through simple to complex problems completely. Decision makers will be able to cut to the heart of important relationships that connect "points of wisdom" so that optimal decision making will be the order of the day. Also future computer technology has the potential to assist in the development and implementation of truth management systems as mentioned in Chapter I. Additionally, future corporate technology will have a tremendous impact on changing computer storage, networking, and software development over time.

An Open-Ended Infrastructure Helpful for Optimal Decision Making

A logical starting point for effective computer storage and networking operations is recognizing the importance of making optimal decisions for growing a learning organization over time. Basically, this means providing an *open-ended infrastructure*, that is, a common architecture or a common framework so that everyone will be operating off the same page. Some conservative computer personnel dismiss an open-ended approach and employ a wait-and-see attitude. This results in passing up proven, stable, and scalable low cost solutions, such as the Linux operating system, the Apache Web server, the MySQL database, and the Open Office desktop suite. In addition, delaying an open-ended infrastructure can hinder progress since developers are forced to cobble together poor or unwieldy solutions. An open-ended infrastructure depends not only on the opportunity to be explored or the problem to be solved, but also on the maturity of the solution being considered. In the final analysis, companies have to question whether they can afford *not* to support an open-ended infrastructure. An open-ended framework assists all stakeholders in a company's operations, whether it is on the local or global level today or tomorrow. In the next chapter, an open-source model for software will be explored.

Utilization of Business Process Management for Connecting "Points of Wisdom"

Business processes provide the logical flow and context in which data, information, knowledge, and their resulting intelligence and optimization is generated, processed, and analyzed. Effective, well documented, and understood processes are the foundation for all organizations. In terms of business and machine-to-machine processes, work activities tend to take more time when humans are involved, especially if judgment is required to deal with exceptions in a highly repetitive process. So-called "long-running processes" like credit approval, new product development, and clinical drug trials involve many starts, stops, and detours that traditional workflow systems are designed to handle. Workflow systems typically assign roles to the people in the process, and use rules to determine the flow of work to those people. Essentially, the job of the workflow systems is to act like a very large engine. For example, CBCA, an Oakland-based claims-processing outsourcer for self-insured employers, is deploying workflow to help its 60 claims examiners handle 15,000 claims per day more efficiently. It allows the company to separate the problems into separate blocks. The workflow system, part of BEA's Business Process Manager, divides the tasks into queues that can be handled as appropriate by either machines or humans. It is not eliminating humans; it is only having the human element undertake the more difficult problems. The net result is that the system is handling more claims with fewer examiners, better consistency, fewer errors, and enables the company to offer more complex products (Margulius, 2002).

As enterprise business processes become more automated and more interconnected with outside partners and customers, workflow systems play a larger role in the world of *business process management* (BPM). BPM vendors include IBM, Microsoft, Oracle, and BES. Additionally, some BPM vendors, such as Tibco, WebMethods, and BEA Systems, have acquired workflow companies including InConcert, Intelliframe, and Workflow Automation. Business process management allows organizations to analyze workflows and initiate best practices to manufacture, market, deliver, and support their products and services in the most efficient manner possible for better bottom-line performance. Through BPM, organizations can streamline, automate, and optimize their business processes to continuously improve and enhance business performance for maximum competitive advantage. BPM allows a company to look at its business processes, then figure out where technology can add value, that is, find a better way about what needs to be done organizationally over time. Overall, effective business process management ensures that a company's people and computer technology work together as a cohesive team.

Within a BPM operating mode, business processes for a company's functional area provide the logical flow and context in which data, information, and knowledge is generated, processed, and analyzed for business intelligence and optimization for a company's operations. Essentially, the foundation of BPM is effective, well-documented and understood business processes. As such, BPM provides the means for connecting "points of wisdom" to assist decision makers in taking a comprehensive approach to organizational operations today and tomorrow. As shown in Figure 1, corporate planning activities can be connected

Figure 1. Relationship of corporate planning to a company's functional areas (for connecting "points of wisdom" within an optimal KM/WM system environment)

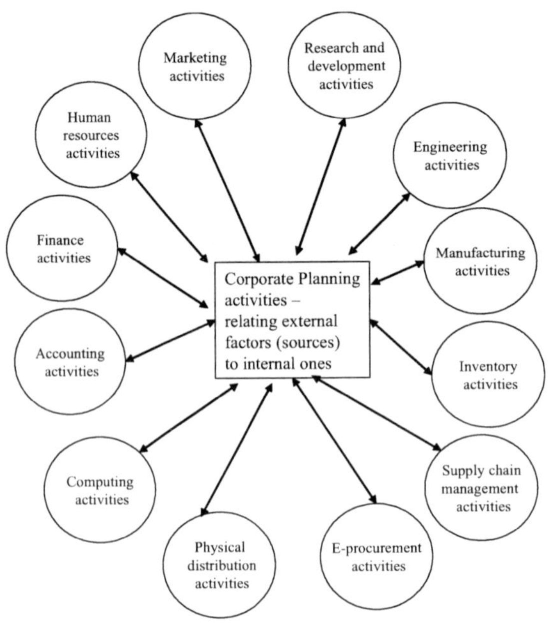

to external and internal factors (sources) that are related to an organization's functional areas. In turn, these functional areas can be designed such that their business processes complement the ability of decision makers to make sound decisions daily as well as over time. The net result is that the big picture can be pursued for a holistic approach to organizational operations.

Employment of Information Lifecycle Management to Oversee Computer Storage Growth

To assist decision makers in utilizing optimal KM/WM systems, it is helpful to employ *information lifecycle management* (ILM) to oversee computer storage growth. Essentially, ILM is a set of management and operational practices that align the business value of information, knowledge, and resulting intelligence and optimization to their most appropriate and cost effective infrastructure. What is important about this approach is that implementing ILM is not as simple as plugging a software solution in to an existing infrastructure. But rather, ILM relies extensively on processes and the people who execute them.

By approaching storage intelligently, application awareness can provide a foundation for information lifecycle management, a strategy that is widely touted as the future of storage. ILM aims to define a set of policies and automated processes for provisioning, mirroring, replicating, snapshoting data migration, and retention based on the value of different types of information to the organization. ILM goes beyond the application, thereby allowing policies based on awareness of data itself and even the information, knowledge, intelligence, and optimization that data represents. It is expected that the ILM initiative will continue to develop further definitions, reference models, and educational materials that will help customers and vendors implement ILM without being tied to any specific solutions or technologies.

In a later section of the chapter, there is a discussion of storage virtualization. Storage virtualization is a way to facilitate information lifecycle management, the movement of files between expensive disk arrays and low-cost media for data archiving, and to shield applications from underlying physical-storage systems. It combines storage systems from multiple vendors into a single storage pool for greater utilization of disk arrays. The bottom line is that storage virtualization enhances the ability to perform information lifecycle management.

Computer Storage Management on the World Wide Web to Facilitate Decision Making

An approach to realizing an effective optimal KM/WM system-operating mode is related to computer storage management (CSM) on the World Wide Web. Organizations increasingly want their business applications and infrastructures to operate in a Web-based or an intranet-based environment. Web-based CSM allows decision makers in any location to use information, knowledge, intelligence, and optimization results and then collaborate on their results and meaning in terms of the total organization. Clearly, a Web-based CSM approach offers attractive benefits for organizations. With platform-independent Web clients, decision makers can access facts and figures from any computer in any location and can collaborate in business processes. With three-tier architectures, Web-based computer storage systems can adequately support large numbers of decision makers accessing these systems through Web servers. Browser-based clients typically require no user training or special expertise, and they do not require extensive client-side administration and maintenance. Going a step further, four-tier architectures can be useful for assisting decision makers.

To assess Web-based data management functionality, these systems should utilize solid back-end architectures and approaches designed to optimize the Web, including thin clients, three-tier architectures, Web-based administration, and support for standards such as HTML, Java, ActiveX, and XML. As the Web continues to make inroads, organizations have accepted the Web not only as a data, information, and knowledge distribution medium, but also as a platform for actionable applications based on intelligence and optimization results. Overall, computer storage management on the Web provides companies with innovative ways to leverage their business resources and processes for competitive advantage today and tomorrow.

Enterprise Portals Found in Effective Computer Storage Management

An enterprise portal (EP), also called a corporate portal (CP) or an enterprise information portal (EIP), is a window into the myriad of data streams, applications, networks, and processes that make up an enterprise. It is a layer of technology that unifies the data, information, knowledge, and resulting intelligence and optimization found in an organization via a single point of access. For the most part, the synthesis of one, two, or three of the myriad transactions occurring during the course of a day could have incredible value if they are shared. The convergence of those transactions and their resulting intelligence streams could result in some insight that, shared through an enterprise portal, could give decision makers an insight not had previously so that better and wiser decisions can be made.

An enterprise portal on the public Internet employs a profile of the user's information requirements and the services of a search engine to help the user find what is wanted quickly. An Internet portal provides the user with a single interface to the vast network of servers that constitute the Internet. A typical enterprise portal, on the other hand, in the corporate environment has a similar objective: to provide business users with a single interface to data, information, and knowledge scattered throughout the enterprise for the purpose of distilling business intelligence and optimization results. Such enterprise portals fall into two main categories. First, a *collaborative processing enterprise portal* helps users organize and share workgroup items, such as e-mail, discussion group materials, report memos, and meeting minutes. Second, a *decision processing enterprise portal* helps managers and business analysts access corporate data, information, and knowledge as well as their resulting intelligence and optimization results for making wise business decisions. This type enterprise portal supports a wide range of different types of corporate business facts and figures and offers significant potential for growing optimal KM/WM systems. Three leading portal products are Plumtree Enterprise Portal, IBM Websphere Portal, and SAP Enterprise Portal.

In the e-business world, enterprise portals have become a permanent fixture. As an example, New York City-based e-Steel Exchange (www.esteel.com) is a steel industry portal that offers industry news, provides decision-making tools, and is an electronic exchange for both buyers and sellers. E-Steel uses various products, where Yantra's PureEcommerce Supplier Portal is the most prominent portal tool. In addition, companies are using enterprise portals to drive supply chain efficiency by minimizing shortages and improving order management and fulfillment, thereby enhancing relationships with trading partners. This supply chain approach promotes new relationships such that suppliers want to do a better job. Since enterprise portals have been integrated with Web technology from the start, there is a natural fit with wireless access. Wireless access capabilities are being integrated with general portal access capabilities rather than directly into data warehouses or business applications. Newer enterprise portals will use computer-networking technology that is tied-in with wireless access capabilities.

Development of a Corporate-Wide Computer Storage Infrastructure

In the development and maintenance of an optimal KM/WM system, a most important factor relates to its computer storage infrastructure. Because of the ability to get at vast amounts of facts and figures plus their results provides decision makers with an important competitive edge, a company needs to share these results across the organization. Many times layers of management can be eliminated and the organizational structure can be flattened. These flattened organizations allow decision maker to make and implement decisions quickly. Sharing data, information, and knowledge along with the resulting business intelligence and optimization results with customers and suppliers helps break down the organizational boundaries. Allowing people to share and work together results in enhanced teamwork, and when employees are empowered to deal directly with their customers, the quality of service can improve by cementing relationships with the customers not found in the past.

Capturing an organization's important facts and figures and making them accessible to decision makers across the company requires an investment of time, money, and talent. The right systems must be in place and the right applications chosen for accessing computer storage. In addition, the company's employees must learn to understand and use the appropriate database tools. Pivotal issues involved in providing the right database technology are complex. For example, in today's newer structures, data, information, knowledge along with resultant business intelligence and optimization results are closer to the individual. This means that the centralized information system organizations are no longer the clearinghouses, but rather provide guidance and standards while the line organizations of individual business units own their systems.

Overall, the development of an appropriate computer storage infrastructure within an optimal KM/WM system environment can reduce costs and increase profits through the reorganization, that is, reengineering, of business processes and functions. The new computer storage infrastructure should also facilitate the development of alliances with customers and trading parties. These relationships should provide a wider range of services than anyone of them can provide on their own. In the development of an appropriate infrastructure, data marts, data warehouses, very large databases, knowledge bases, data federation systems, an enterprise storage network, storage visualization, and RAID technology are employed. In turn, these computer storage approaches can be related to the need for backing up data and guarding against its loss. All of these important areas are discussed next.

Data Marts, Data Warehouses, and Very Large Databases

A data mart, which is a subset of an enterprise data warehouse, performs the role of a departmental, regional, or functional data warehouse. Typically, an organization builds a series of data marts over time and eventually links them via an enterprise-wide, logical data warehouse. On the other hand, a data warehouse contains all the metadata, business rules, and processing logic required to scrub, organize, package, and preprocess the data. It contains the information required to find and access the actual data, wherever it actually resides. An important element in both database approaches is the level of complexity and size. The data model for a data warehouse normally increases in accordance with the number of lines of business it serves. In contrast, a data mart, regardless of size, generally focuses on a single subject area. A data mart is generally determined by the department or line of business for which it is created. Initially, data marts that an organization develops are marketing and sales marts since the customer information they contain is essential to its business. Data marts are typically subdivided into independent and dependent. An independent data mart derives its data from a number of sources and operates autonomously while a dependent data mart is essentially a subset of the data warehouse. The nature of the organization's business drives whether a data mart will be independent or dependent.

Today, data warehouses can be looked upon as a place where data is copied, transformed, and stored multiple times in order to meet usability, performance, availability, security, consistency, and other needs. This process requires rigorous analysis and modeling of the data environment and the use of extract, transform, and load (ETL) tools. Originally, the data in data warehouses was extracted from operational systems and loaded to a warehouse in order to protect the operational systems from the vagaries of user queries. Doing so also made it possible to combine and cleanse the data and create the new data elements users required. Defining data marts as a separate layer made it possible to optimize them for a variety of specific user needs; the business data warehouse became the single system for enterprise information. Then came the need for data to flow more quickly through the warehouse and, in some cases, for bi-directional data flows to better support CRM (customer relationship management), call centers, and e-business. In order to meet these needs, the operational data store developed as a separate layer.

The current data warehouse architecture is based on the principle of moving the data to the query. The opposite approach, called distributed query, is to move the query to the data. Distributed query enables direct, real-time access to the original data on distributed and disparate sources. It also eliminates the need to maintain multiple copies. However, a distributed query makes query performance less certain, prevents any required preprocessing of data, and introduces the possibility of sending a "query from hell" to the operating environment. By using the distributed query approach in specific, well-defined circumstances within a traditional data warehouse provides some new possibilities for addressing users' data needs. This distributed query functionality is available in the IBM DB2 Information Integrator.

Currently, companies are shaping their decision-making capabilities with real-time data warehousing. More specifically, companies are consolidating their sales and related data onto a centralized, enterprise wide data warehouse. Consolidating sales information from several different customer-facing channels enables the company's decision makers to have real-time access to vital sales information, thereby allowing them to quickly analyze customer buying patterns. This analysis allows them to create comprehensive sales plans and forecasts for stocking the most desirable products for its customers. In essence, a company's decision makers have the ability to judge soundly about what needs to be done to better match stocked items with customer requests over time.

As companies grow and their systems expand to meet anticipated growth, they are coming up against the limits of their current system architectures, in particular, relational database management system (RDBMS) software, hardware platforms, and manageability. Information systems, which have historically been kept small to achieve maximum transaction rates, have raised their ceiling limits. For example, automated teller machines today offer customers access not only to savings and checking accounts, but also to a full array of financial services. The net result is that improved service requirements demand more information, knowledge, and resultant intelligence and optimization results about customers and products be placed online to assist sales and support personnel. In addition, as users demand more from these kinds of systems, these demands are driving up the sizes of computer storage and their related databases to what is currently called *very large databases* (VLDBs).

Future VLDBs will differ from those of the past not only in their size and the platforms on which they run, but also the framework with which their developers undertake their construction. Database developers are even beginning to plan for multipetabyte databases where a petabyte is defined as 1,000 terabytes. It is predicted that it will not take as long to go from terabytes to petabytes as it did from gigabytes to terabytes. Typically, very large databases for optimal KM/WM systems are simply going to be quite larger than their predecessors for housing appropriate facts and figures. The larger vendors today are Compaq, EMC, Hitachi Data Systems, and IBM where customers want vendors to deliver systems that are more scalable, easier to use, and less expensive.

Knowledge Bases

Related to very large databases are expert systems or knowledge base systems. Expert systems make use of *knowledge bases*, which consist of IF-THEN rules, mathematical formulas, or some knowledge representation structure to represent the knowledge of experts in a certain domain. They scan through their knowledge bases to find the appropriate rules, formulas, or some other knowledge structure to apply. The power of these systems derives from the knowledge they possess, rather than from the inference mechanism employed. The acquisition of expert knowledge becomes a vital task in their development process along with their maintenance over time. Because knowledge in some cases is being changed on a daily basis where knowledge can be derived from many sources, maintenance is an important issue. Also, there are cases where some of the knowledge is not documented and not well distributed. To assist in the maintenance of knowledge bases, it is essential to establish detailed plans that include extraction, analysis, and verification. This approach can assist a company's decision makers in assuring that its knowledge bases are current and relevant to the times.

In assisting decision makers at all levels of an organization, there is generally a need for multiple knowledge bases along with databases, including data marts and data warehouses, to be integrated into a corporate-wide knowledge base. This integration into a corporate-wide knowledge base is necessary in order to have a computer storage infrastructure that supports decision makers throughout the organization. The integration of appropriate data, information, and knowledge no matter their source, including a company's functional areas (i.e., corporate planning, marketing, manufacturing, accounting, and human resources), allows decision makers to find what they want in one place. Overall, multiple knowledge bases are not an island unto themselves but are an integral part of an organization's computer storage infrastructure.

Data Federation Systems

Just as the federal government divides power among parties or states, a federated data architecture, which underlies data federation systems, consists of multiple, distributed data stores with varying degrees of affinity. A federated data architecture provides users with a unified view of distributed data sources. The federated architecture may include data marts, data warehouses, transaction processing systems, and external data repositories. In some cases, these architectures may also integrate XML content, file systems, document databases, and other sources of unstructured content. From this broad view, a federated data architecture consists of multiple, distributed data stores with varying degrees of affinity, thereby allowing for global data access but local data autonomy. Data federation is about enabling middleware to reach out and touch data from a variety of sources, then manage it as if it were in one relational database. A data federation approach allows a company to migrate data from a variety of sources, such as legacy and nonrelational systems, into a single repository.

As such, an effective data federation system integrates various independent data systems by combining aged and real-time data while leaving the source data in place. The user can query (and update) a collection of databases as if all data were at a single location. Essentially, the underlying structure of a data federation system is one that integrates multiple physical databases into a single logical one, letting locally controlled systems act with a global perspective. By using data federation within an optimal KM/WM system-operating mode, companies can achieve the maximum amount of architecture possible within an organization's realities by sharing critical metrics and measures across its business operations.

Enterprise Storage Network (ESN), Storage Area Networks (SANs), and Network Attached Storage (NAS)

A company's computer storage infrastructure can also be viewed from an *enterprise storage network* (ESN) perspective within an optimal KM/WM system-operating mode. Essentially, the focus is on improving access to the large volumes of data residing in disparate storage systems. ESN is an architecture used to extend the reach and flexibility of a data infrastructure in order to leverage its value to decision makers in an organization. As such, an ESN architecture consists of both enterprise storage and storage networking technologies to deliver a common way to manage, protect, and share information knowledge, intelligence, and optimization results regardless of distance or scale to decision makers. In the design and implementation of an enterprise storage network, there is need to take into account how much growth, the number of connections, information security, how it will scale, availability, performance, and the services required. Today ESN represents a strategy that includes *storage area networks* (SANs), *network-attached storage* (NAS), and direct attached connections because the needs of most practical situations cannot be met by a single connection topology. ESN encompasses both SANs and NAS in order to address the demands of realistic storage environments.

Essentially, SANs provide high-speed network storage that is external from processing servers. By separating a SAN from the server, performance is greatly improved since data storage and access tasks are removed. Generally, SANs connect with Fibre Channel, which allows them to share access to multiple storage devices such as tape or RAID (redundant array of independent/ inexpensive disks) system. In the past, storage devices were attached to the network via an available server. This limited the data that could be housed on a particular storage device. Since the server would run a designated operating system, the storage device that was attached to that server would only be able to store information from that operating system. In contrast, a SAN environment changes that concept entirely since multiple servers with different operating systems can share the same storage devices.

Complimentary to SANs is network attached storage, which allows for a complete, enterprise-wide storage system. NAS provides distributed storage for workgroups and departments by helping companies avoid having dozens of servers to administer and maintain. Because NAS does not get rid of all administration costs network administrators still have to set up work groups and assign permissions and passwords for each NAS device. If there are more workers accessing a NAS device than reside on its subnet, a storage area network with multiple clustered servers and storage devices can be built.

It should be noted that SANs differ from NAS in that they are composed of storage devices and servers connected via high-speed network connections, usually Fibre Channel or possibly gigabit Ethernet. SANs can be server hosted where the software needed to operate the storage devices resides on one or more of the servers or they can be peer-to-peer, where each storage device on the network manages itself with its own thin server. Since SANs are scaleable and extensible, they are solutions for companies that need large, centralized storage repositories for active archives or backups. In contrast, NAS is for distributed storage for workgroups and SANs are best for centralized enterprise wide use. Both complementary technologies have a place within an optimal KM/WM system. The power, reliability, and extensibility of storage area networks are also of great importance to implement e-commerce in every industry. However, the level of complexity typical of e-commerce and SAN systems result in major challenges (Erlanger, 2004).

Storage Virtualization

The ability to centrally manage storage and make it available to multiple machines is known as "virtualized storage" which has been implemented on computer mainframes for years. Today, the term *storage virtualization* means creating a "virtual pool" of storage devices that can be accessed and managed from a central location. True storage virtualization enables database administrators the ability to carve out space, expand storage, link to storage on other devices, and logically move terabytes of data in seconds. Basically, the user uses a single interface that provides a logical view rather than a physical configuration of the storage devices. With current virtualization techniques implemented in software and sometimes in hardware, the user does not need to know how storage devices are configured, where they are located, their physical geometry, or what their capacity limit is. The advantage of separating the logical and physical storage devices center on allowing the business application to access the logical image while minimizing any potential differences of the underlying device or storage subsystems. Physical storage devices can be added, upgraded, or replaced without disrupting application or server availability by separating logical and physical characteristics.

Along with SANs and NAS, storage virtualization is one of the compelling newer trends in the storage networking industry. Storage virtualization enables storage pool and device independence, which creates a single point of management from what were previously many (diverse) points. Since storage virtualization of SANs and NAS can simplify the administration of a very complex environment, there is also the ability to improve interoperability in the absence of open-storage networking standards. Storage virtualization in today's open-ended systems environments means much more than the centralized view and management of storage pools in a heterogeneous environment. It also means improving storage manageability.

Secure Data Using RAID Technology

RAID (Redundant Arrays of Independent/Inexpensive Disks) was originally described as a way to combine multiple disks and virtualize them to the operating system as a single disk. Related to this way of describing RAID is the storage of the same data in different places redundantly on multiple hard disks and protecting it against loss. This can also be accomplished by using magnetic tapes for storage of data in different locations and guarding against its loss. As developers constantly create new and more advanced ways of managing and controlling data along with information, knowledge, intelligence, and optimization results, RAID has become increasingly popular. The amount of data being stored on hard drives has increased exponentially, along with the need to keep that data accessible. With that need in mind, manufacturers have developed and provided various different solutions to the ever-expanding market.

Developed by the University of California-Berkeley computer sciences department about twenty years ago, RAID was envisioned as a much-needed source of storing data due to changing computing requirements. Since the computer became a critical part of an organization's business, more and more data was stored in the computer and this data along with information, knowledge, intelligence, and optimization results was needed to be available twenty fours a day, seven days a week. Thus, several challenges surfaced in the storage industry, one of which included the need to match system I/O performance to the pace of computing performance so that access to data did not become a limiting factor for applications. Also, there was a need to provide uninterrupted access to online data at levels of reliability in excess of the expected lifetimes of the computer systems that processed it. Today and in the future, computer storage and security will continue to be one of the most important technologies to the growth of businesses. Typically, RAID solutions will enable companies to manage and control their important data, including information, knowledge, intelligence, and optimization results. As various types of storage solutions are developed, business organizations will be provided with greater data reliability and accessibility for developing and growing an optimal KM/WM system.

Keeping Computer Storage Safe and Secure Against Intruders

In the past, most information system and data center managers utilized a *firewall* on a gateway to keep computer storage safe and secure, as well as locking up equipment to keep the intruders out. But with the proliferation of technology and the use of a broad range of applications and devices, it is not enough anymore to tweak a few settings and know where the closet keys are kept. In addition, companies need to shore up *password* security by purging passwords that are no longer used but do allow access to computer storage. Where deemed necessary, a company should encrypt its data and related information with backup. Although *encryption* is not hassle free, a company that builds encryption into its security plan will see benefits sometimes down the road.

Because e-commerce customers trust their personal data only to online businesses that promise to guard those digital assets, smart companies realize this and are taking steps to protect their customers' data assets. Many times, e-businesses are so eager to get up and running that they skip some key steps in data security. The development process needs to include measures to detect hackers, spammers, and virus writers as well as to educate employees about security. Business-to-consumer (B2C) enterprises need to look at privacy, confidentiality, and ID theft while business to business (B2B) operations should think about a much higher level of authentication (because money and intellectual property are involved) and granular authorization (providing customers, partners, suppliers, and employees with selected access to data content and applications).

In terms of B2C and B2B operations, however, the way that intruders get into enterprises is broadening. They still use the tried-and-true method of delivering executable files via e-mail, but they are also targeting devices and applications that have not been used much for virus delivery in the past. Recently, IBM Global Services unveiled its Business Security Index report, which noted that mobile devices such as cellular phones and PDAs will be under increasing attack from hackers, spammers, and virus writers. The company also predicts that embedded car devices and satellite communications systems will become targets as malicious intruders look to do the widest possible damage with the least amount of effort.

Going beyond the attacks by hackers, spammers, and virus writers, there is a newer intruder on the business scene, that is, *phishing*. Unlike viruses and worms, which tend to be written by bored programmers or security researchers, phishing attacks are undertaken for money. Phishers spoof a Web site or employ other tactics to trick users into giving up their personal information, such as bank account numbers and passwords. A particular disturbing trend in phishing attacks has been the use of tactics that require no user action. Typically, a user will receive an e-mail and have to launch a program or visit a site in order to be duped. But now phishers are finding a way to get onto systems without users knowing about it.

In addition to the above security measures, there should be consideration for *lifecycles and retention* of computer data storage. Today, companies are keeping too much data of all types beyond what is legally required, thereby leaving the data and related information open to hackers. To determine the proper level of retention, the question can be asked, "How long should the data be retained?" By keeping data beyond what is required, it becomes a liability for any company. This dead-in-place storage needs to be disposed of in a judicious manner.

Today, a typical company needs to develop a holistic approach to database administration in order to protect itself against intruders. It needs to leverage the appropriate tools, up-to-date policies, as a well-trained staff for optimal security. Regardless of the tools used to ensure security, there are recommended steps that should be undertaken. Initially, there should not be a sole reliance on technology but rather on management control and intervention. A typical company should align security policies with business strategies since these policies drive company programs. There is need to evaluate what level of security is required and determine whether it adds value to the organization. Although keeping security management in-house will give a company full control, finding and training quality security personnel in a tight labor market can be difficult. Additionally, outsourcing security is an option. However, there is a need to get a strong service-level agreement to compensate for the lack of control over out-of-house personnel within the company as well as outside its boundaries.

Development of Corporate-Wide Strategic Computer Networking

To implement an optimal KM/WM system, a corporate-wide computer networking plan starts with defining the services that the computer network will offer to decision makers for optimizing operations daily and over time. Services mean the network application protocols the network will support, like IP (Internet protocol) and Systems Network Architecture (SNA). Most large companies utilize a large number of application protocols, but companies are reducing that number to simplify support and reduce capital equipment investments. If a company has only one protocol (usually IP), it is ahead of the game. As the growth of IP-based traffic continues to expand, it has placed new demands on network traffic patterns. Network management and quality of service are vital to successful VOIP (Voice Over Internet Protocol) network implementation. VOIP is the true convergence of data, voice, and video over a single Internet Protocol infrastructure. VOIP was created to consolidate the need for multiple interfaces to retrieve information from multiple mediums (Travis, 2004).

In addition, another important technology of late is wireless broadcast technology (WiFi). WiFi allows digital devices within a several-hundred-foot range to connect at broadband speeds to the Internet. It is already used in millions of homes in the United States to create wireless home networks. Thousands of public-access points are popping up worldwide, especially in coffee shops, airports, and hotels. Every Dell and Apple laptop now ships with WiFi capability. The next step is combining VOIP with WiFi. That could be a significant improvement over the still-expensive cell phone networks. When WiFi technology is combined with a VOIP network, the net effect is that users have the ability to make wireless calls over the Internet. For providers of VOIP services, because calls are routed through the Internet rather than over telephone lines or cellular networks, infrastructure costs are much lower. Overall, the implementation of newer corporate-wide computer networking capabilities means that connecting points of wisdom for decision makers will be made easier.

Well-Planned Content Management Strategically Distributes Information, Knowledge, Intelligence, and Optimization Results

Related to computer networking is *content management*. It can be defined as the end-to-end process of acquisition; collaboration within the enterprise to be able to manage the content; and then the dissemination and billing for that content to the Web site, to partners, to channels, to syndication, and to other organizations.

Although content management's roots in the past lie in document management, its future will likely lie on the World Wide Web and beyond as its evolution pushes the concept of what content is and how it can be used for e-business. Content management was given a major boost by the Web as companies began to realize that running business operations on the Web has several important benefits. Among them is the fact that companies are finding a need to collaborate around content, and that often means bringing together users and content from local users as well as from different parts of the world.

With content management taking more and more of a role in business, it must be extended to include more content sources and access methods, as well as different kinds of content. Although content itself is still important, it cannot be seen in isolation where it is reduced to just text and graphics. But rather, users must be able to manipulate those kinds of content online. Content management cannot be limited to Web content management, but must cover the entire enterprise. From this view, it can be called *enterprise content management* where the user can bring in the Web content, the documents that are inside and outside the company, into one content management infrastructure that is both local and global in nature and distributed in structure. Essentially, then a well-planned enterprise content management strategically distributes information, knowledge, intelligence, and optimization results within the company and outside the company via the Web within an optimal KM/WM system (Sanborn, 2001).

Use of Computer Networking with Great Accent on E-Commerce

In a typical organization today for computer networking, a creative approach is needed to merge e-commerce with the Internet, intranets, and extranets. The ingredients include creative ideas, business strategies, value propositions, newer computer techniques, and the latest e-commerce technology, among others. The real winners of tomorrow are those that are not afraid to experiment with e-commerce technology. Experimenting means challenging the traditional e-commerce assumptions and going against the grain (that is, being a revolutionary and believing in one's ideas when others are not convinced). This could include paying one's customers and perhaps even one's competitors, changing services, and providing value in new ways. Hence, there may be a need to reshape traditional business models and develop new e-business models that are emerging and figure out how to leverage or improve them. Being in a highly changeable world where customers can be competitors and competitors are customers, where new profit centers can be created out of traditional cost centers as in the case of Internet bill presentation and payment, and where traditional power structures and businesses are being overturned and transformed by newer approaches, the best e-commerce approach begins with a creative idea that is quite different from the past.

Focus on Wireless and Mobile Technology

A recent direction in the Internet is *wireless* access whereby people anywhere feel like they are connected directly to the Internet, without a wired connection, and without requiring dial-up access. This wireless approach to the Internet opens doors to applications that people have begun to imagine, especially when trying to connect points of wisdom within an optimal KM/WM system. Motorola, for example, has embarked on an ambitious and expensive plan to develop a global wireless Internet access system for use in conjunction with third-generation cellular networks. The company will invest upwards of $1 billion over the next ten years, much of that directed toward its technology partner, Sun Microsystem Inc., which will provide the software and hardware. The project is an extension of existing digital wireless networks, offering broadband network access to the masses with service comparable to current corporate LANs.

In the future, the total number of wireless devices is expected to reach into the trillions, with the billion-plus cellular phones out there representing only a small piece of the total. The bulk of wireless devices is expected to be sensors and monitors, working in a machine-to-machine mode and driven by requirements for stability, safety, and cost efficiencies. They will monitor temperatures, stress, access, vibrations, sound, dust and air particles, and a seemingly endless parade of parameters. This is where self-forming, self-healing, peer-oriented capabilities of ad hoc networks really begin to shine. Additionally, wireless devices will work collaboratively to gather and analyze data, working autonomously until the collective has reached a conclusion about their data, and then reporting the result to another node or system. This peer-to-peer versus hierarchical approach allows nearly infinite scaling to be possible.

The current impetus to outfit more front-line employees with *mobile technology* is not new. UPS, for example, actually built its own networks for monitoring package delivery before mobile telephony was available. IBM was an early leader in developing networks for field salespeople. Worldwide, the trucking industry and other logistics businesses have used mobile technology in the field for many years. It should be noted that top managers are more likely to value and support any mobile solution if they themselves are mobile. For example, at the Canadian network-equipment maker, Nortel Networks, 12,000 of the company's 35,000 employees are "teleworkers" who spend a significant part of their workweek on the road or in home offices. Essentially, work is an activity, not a location (Karaian, 2003).

Using Blogging to Assist Decision Makers in Connecting "Points of Wisdom"

Today, blogs are the soapboxes of the Internet, that is, they represent everything from personal diatribes to political discourse to technical gadget reviews. However, with the growing popularity, enterprising organizations are using blogging to engage in an open dialogue and exchange of ideas with present customers and future ones. Not only should good comments be welcomed, but also those that are bad or derogatory toward the company as well as its products and services. In this manner, there is a natural extension of communications with customers. Similarly, internal comments about a company's operations should be encouraged so that management can rectify a bad situation with its employees. Whether blogs are received externally or internally, a company's decision makers can use them to connect important points of wisdom that cut to the heart of the relationship concerning future opportunities or resolving pressing company problems. In addition, blogging is an effective way for a business to keep in touch with its community as well as share ideas, gripes, and news. It gives a human face to what the public might perceive as a faceless corporation. Overall, blogging can be extremely helpful to a company's decision makers about getting comments about its products and services as well as what is happening internally and externally about its operations in order to connect points of wisdom.

Optimization Software for Wired and Wireless Technology

To help companies manage their networks more easily and less expensively, a number of software vendors are providing this capability. Their focus is helping companies get a better handle on whether they are making optimal use of their network and application resources. The Flame Thrower product family from Antara.net, for example, is software designed to optimize network capacity and test how network devices will perform in high volumes of traffic. Other software vendors and their products include the following. First, enterprise Perspective from Keynote manages Web application performance on either side of the firewall and displays results on existing management consoles. Second, nGenius Capacity Planner 3.0 from NetScout is useful for long-term network planning, managing application service levels from the end-user perspective, and customizing performance reports. Third, FastNet from NetTest is network hardware with embedded software that increases the speed and quantity of traffic on a network. Fourth, RiskWise product suite from Parc Technologies manages e-business networks and provides traffic analysis, resilience analysis, and topology optimization (Dubic & Fontana, 2001).

Recently, too many users were experiencing long delays when using wireless and mobile technology. Field representatives for a water utility, for example, spent part of their days waiting for the sluggish Cellular Digital Packed Data (CDPD) modems in their laptops and PDAs to transmit and retrieve information. When the system took up to four minutes to transmit a single completed service ticket, network managers knew that they needed a solution to move the wireless data much faster. A number of vendors, including startups BlueKite, BroadCloud, Fourelle, Idetic, and XOSoft, as well as the more established provider Inktomi responded by developing *data optimization software*. This newer software speeds up data transmission over existing cellular wireless networks although it is somewhat expensive. But for some businesses, the time saved is worth the investment. Wireless carriers like Sprint PCS, Verizon Wireless, and Britain's BT Cellnet are investing and continue to invest in the optimization packages and passing on the performance gains and required investment costs to their customers and trading partners.

Typically, sending data over a cellular wireless network can be traced back to a single source, that is, wired technology pressed into service in a wireless world. Specially designed wireless applications and protocols like the Wireless Application Protocol (WAP) are built to manage the slow speeds and unpredictability of radio-based data transmissions. Unfortunately, many companies are running applications and network protocols designed for the relatively high bandwidth and consistent connections available over fixed wires. In contrast, optimization systems smooth out these differences by placing a small software client on the wireless device and a server on the other side of the wireless connection. When linked together, the client and the server act as a conduit for data flowing to and from the wireless device, compressing the data, removing extraneous information, and simply maintaining the link between the wireless device and the network. Currently, data optimization software is the only real answer to solving data transmission speed and reliability problems. It allows businesses that need complex data delivered speedily to do so without needless waiting (Bruzzese, 2001).

World Wide Web Globalization as It Affects Optimal Decision Making

Due to worldwide globalization, U.S. companies who are finding operating their e-businesses as though their country is the only one, especially since more than half of the world's online users do not live in the United States, are in for a rude awakening. These global users speak different languages, use different currencies, abide by different legal systems, and have very different preferences, experiences, and tastes than their American counterparts. Although the Internet has opened foreign markets that have been costly and difficult to enter, a simple Web site can attract visitors from all over the world. A Web site written in English, marketed with an American mentality, and priced in U.S. dollars only will have a difficult time turning visitors into buyers. Because the globalization process is more complex than simply translating an existing English site into another language, companies need to start with the market itself and its challenges and go through those challenges one by one to figure out what they need to do today and tomorrow.

To help global e-businesses manage, synchronize, and update the content of multiple-language sites, there is a new breed of software. This global Web software includes translation into one or more foreign languages and Web-site localization that helps companies adapt their businesses to a particular market or culture. In addition, companies need to consider cultural and political aspects when taking their business abroad. Other factors that need attention include legal, staffing, competitive, and payment issues. This is extremely important since 75% of the world's Web Market is expected to live outside the United States in the next several years. Hence, the message is clear about what needs to be done (Gareiss, 2000).

Due to globalization for selling goods and services, it is helpful for decision makers to utilize optimal decision-making techniques and related software to help in developing optimal KM/WM systems as well as growing them. That is, decision makers need to stretch their view of the world beyond the United States. When considering where to place sales marketing centers, manufacturing plants, maintenance service centers, and the like, the many facets of each major country of the world needs to be formulated. In this manner, appropriate company facilities that tie-in with their trading partners can be determined in the world. The net result is that all parties can reap the financial benefits of optimization on a global basis, that is, the right hand knows what the left hand is doing. Participating companies are not operating in a vacuum but their decision makers can benefit fully from collaboration using new computer technology and processes that are related to an optimal KM/WM system.

Summary

Initially, computer technology that will assist in developing optimal KM/WM systems today and in the future was explored. Next, a discussion of an open-ended infrastructure helpful for optimal decision making and business process management that is useful in tying together the various points of wisdom for making sound judgments by decision makers were set forth. Additionally, the employment of the information lifecycle management to oversee computer storage management was presented. Effective *computer storage* as found in optimal KM/WM systems was approached from the standpoint of facilitating optimization of a company's operations. The development of a corporate-wide computer storage infrastructure using data marts, data warehouses, very large databases, knowledge bases, data federation systems, an enterprise storage network approach, and storage visualization was set forth. Also, this part of the chapter looked at keeping computer storage safe and secure against intruders. In the next part, important concepts underlying *computer networking* were discussed. Underlying such a networking infrastructure are the Internet and the World Wide Web along with intranets and extranets. In addition, the current focus on wireless and mobile technology was detailed, along with blogging. Lastly, worldwide globalization as it affects decision makers in optimal KM/WM systems was examined. Overall, computer managers need to think in terms of how their companies tie in with global concerns today and tomorrow when assessing a company's computer storage and computer networking needs.

References

Bruzzese, S. (2001, September). No Need to Wait for 3G. *MBusiness*, 55-58.

Dubie, D., & Fontana, J. (2001, May 7). Make way for new management tools. *Network World*, 10.

Erlanger, L. (2004, October 25). Building smarter storage. *InfoWorld*, 47-52.

Gareiss, D. (2000, December 11). Business on the World Wide Web. *Information Week*, 69-78.

Gartenberg, M. (2001, October 22). Why ubiquitous computing will supplant the PC. *Computerworld*, 23.

Golsborough, R. (2000, January). In search of intelligence. *Office Solutions*, 52.

Karaian, J. (2003, May). Where wireless works. *CFO*, 81-83.

Malone, R. (2001, November). Wearable computers. *Managing Automation*, 48.

Margulius, D.L. (2002). Workflow Meets BPM. *InfoWorld*, April 22, 64-65.

Pasternak, D., & Norseen, J. (2000, January 3/January 10). Reading your mind and injecting smart thoughts. *U.S. News & World Report*, 67.

Ricodela, A. (2005, February 21). Slow going on the global grid. *Information Week*, 34-38.

Ricodela, A. (2005, April 25). AI's next brain wave. *Information Week*, 45-48.

Rist, O. (2005, May 23). Utilizing computing: A dream deferred. *InfoWorld,* 45-49.

Sanborn, S. (2001, February 19). Content management expands its role. *InfoWorld*, 38.

Travis, P. (2004, December 6). The ease of VOIP. *Information Week*, 65.

Chapter IV

Computer Software Found in Optimal KM/WM Systems

Issues

- To show the importance of computer software to connect "points of wisdom" in a typical company
- To underscore the changing nature of computer software of today and tomorrow that affects optimal KM/WM systems
- To set forth appropriate computer software that is found in implementing and growing optimal KM/WM system
- To present an overview of effective software for optimal KM/WM systems that is capable of assisting decision makers in judging soundly over time

Introduction

In the pursuit of judging soundly about connecting "points of wisdom" in a typical company about what needs to be done today and tomorrow, there is a wide range of current software packages that are helpful in implementing and growing optimal KM/WM systems. This is the subject matter of this chapter. The software explored includes the following: *new business models, optimization, goal programming, product lifecycle management, predictive analytics, and knowledge discovery (data mining).* In addition, *data visualization software* and *virtual reality software* are included. Still other software packages could have been included, such as business intelligence and online analytical processing. In the near future, it is expected that newer software packages will be developed that truly fits under the category of optimal KM/WM systems. In cases where complete optimization is not practical, as in poorly structured problems, near optimum solutions using the above software packages are beneficial to a company's decision makers.

Computer Software Can Assist in Connecting "Points of Wisdom" in Optimal KM/WM Systems

As noted in the prior chapter, a business process management (BPM) approach allows "points of wisdom" to be connected in such a manner that decision makers can take a comprehensive approach to organizational operations today and tomorrow. For example, Sam Walton, the founder of Wal-Mart, connected in a most important way how the customers and suppliers interact. In effect, he connected the "points of wisdom" such that his stores became the enabler of this interaction between customers and suppliers. Basically, this connection of points of wisdom was accomplished by a sophisticated series of software packages that connects Wal-Mart and its customers to its multitude of suppliers. In a similar manner, other companies can connect points of wisdom and their related activities using appropriate software such that decision makers make wise decisions for critical organizational areas.

Enterprise-Wide Open Source Model Useful in Optimal Decision Making

The proliferation of computer software useful in optimal KM/WM systems will be apparent in this chapter. Related to this newer computer software is an *enterprise-wide open source model*. The open source model makes technology available to everyone and anyone — enabling the community to benefit from the creativity of the whole as well as a high quality product that is self-policed by a group of committed contributors. Basically, open source gives control to the customer.

The open source model creates better software by encouraging collaboration such that the best ideas win — not only within a company but also throughout a connected, passionate worldwide community. Users can see the code, change it, and learn from it. Also, software bugs are more quickly found and fixed. As a result, the model often builds more stable, secure, and easily integrated software, generally at a faster pace and lower cost. This transparency means that commercial open source providers must consistently serve customers through high value, performance, ease of integration, and management. If they fail, companies are free to choose another vendor. With open source, technology lock-in and vendor monopolies could become outdated. The open source model can be contrasted with the proprietary model where product development occurs within one company, such a company charges customers to use its software, charges them to fix bugs and problems, and then charges for upgrades. Overall, the open source model frees companies from this cycle and puts the power of choice and innovation back into their hands. In turn, companies are given choices in terms the best software for producing optimal decisions that cut into the real essence of understanding future opportunities and solving current pressing problems.

Nature of Computer Software Changing to Develop and Grow Optimal KM/WM Systems

It should be noted that various types of software useful in developing and growing optimal KM/WM systems will be found outside the typical organization, such as on the World Wide Web and sponsored by software vendors. These software vendors will provide an organization with the capability to access software and related services on demand. An organization will be able to use new applications without installing and tightly integrating them by, in effect, renting — rather than buying — the management skills of a presumably more knowledgeable service provider. This approach will ultimately impact price, security, service levels, and bandwidth. As examples, Microsoft is trying to leverage its desktop hegemony into leadership in the new era with its .Net services platform. Oracle is pushing managed services for its enterprise suite as a way to cut costs and improve the customer experience. Essentially, this approach is expected to apply to all types of software. This could lead to making software bought and metered as needed, just like electricity. The net result is that software and their related support service could be available any time, anywhere, on any device for users throughout the world. "This software on demand" approach could provide the needed impetus to spread the usage of optimal KM/WM systems globally. It is also about giving customer choices in the way they buy, use, maintain, and upgrade software. That flexibility will help define the winners and losers in the software industry (Alster, 2005).

Currently, some software vendors are allowing their users to access source code, thereby giving them *greater transparency* and *visibility* into their products. For example, the primary use for having Windows source code is as a reference mechanism, that is, an encyclopedia, so that when users are creating their own custom applications on top of Windows. They can look at those interfaces and really understand how they work. Currently not too many organizations employ development teams with sufficient expertise to tackle a code base as complex as that of Windows, or indeed, the Linux kernel. Nevertheless, all organizations should have the option of being able to access source code when it is to their benefit to do so (McAllister, 2004).

New Business Models Software

A helpful starting point for developing and implementing optimal KM/WM systems is *new business models*. More so than in the past, new business models are essential for growing a learning organization over time. Typically, a company needs to change its business model that is reflective of the times. The failure of small-to large-sized organizations indicates that their business models are not reflective of current business conditions although optimization techniques and software may abound in the company. What may have gotten a company where it is today may not get it there tomorrow for what needs to be done over time. In light of this fact, many companies are building new business models that are customized for their operations. In contrast, other companies are utilizing packaged software to fit the specifics of their organizations. One such organization is Computer Associates (CA) International, Inc., which is covered below. Also, Computer Associates, along with other companies such as BroadVision and Palisades, market business model software of different types, which are noted below.

Computer Associates' business model enables the firm to be a partner with its customers in maximizing the value captured from CA's solutions. To that end, the company has committed itself not only to providing more flexibility in negotiating contracts, but also to providing ongoing support from a team of CA employees that includes a "customer advocate". A number of organization changes have been undertaken by the company to help deliver the new Business Model to its customers. These include the way salespeople are compensated and rewarded, adding a post-sales support organization, building a new customer relationship organization, and implementing a host of training programs. In addition, there is recognition of revenue ratably over the term of the contract. This accounting treatment has benefits for both Computer Associates and its investors. As CA makes this switch, information will be provided that enables readers of the company's financial reports to make valid comparisons of its operating revenue. The bottom line is that the new Business Model represents the next step in the company's commitment to work closely with its customers as they make and implement critical business decisions. The overall goal is to help them manage a critical, but increasingly complicated, expensive, and evolving part of their own business.

Underlying this new Business Model is the reality of the fast changing times. Over the past three to five years, information technology solutions to emerging areas such as e-commerce, CRM (customer relationship management), and supply chain integration have posed tremendous challenges for CA's customers. Implementation now involves external as well as internal links. The actual solutions are expensive and the underlying technology is rapidly evolving. All of this means that customers increasingly view the payoff from their software investments to be risky. Hence, one way or another, customers are asking for help with the challenges they face in achieving their desired return on software investments (Computer Associates, 2001).

In addition, Computer Associates has data modeling software that is useful for creating and maintaining databases, warehouses, and enterprise data resource models. ERwin, which is a leading data modeling solution, creates and maintains databases, warehouses, and enterprise data resources that allow the user to visualize data structures for organizing, managing, and mitigating the complexities of data, database technologies, and the deployment environment. ERwin also enables the user to link data designs with BPwin for business process modeling and with Paradigm Plus for UML-based application component modeling. ERwin Examiner, a companion tool, is a database validation solution that raises the level of data quality in transactional and data warehouse systems by systematically scanning ERwin models or thousands of lines of SQL code to identify design consistencies. The complete ERwin modeling suite helps the user to simplify the complex aspects of analyzing, designing, and implementing applications and business processes.

Other software that is available from current vendors include reference to the following situation. If a company is trying to model complex business situations, Excel's Solver function gets the user only so far. In contrast, traditional risk using optimization products like @Risk and Evolver take the user a bit further. However, RiskOptimizer, from Palisades (which also markets @Risk and Evolver), handles scheduling, equipment usage, financial, and other problems that cannot readily be crunched with conventional risk-optimization software. For example, it could allow a shop to use staffing costs and its sales records to determine the best business hours. Scheduling studies in RiskOptimizer could also use time-based estimates of traffic jams, rather than simple distance, to plan optimal routes. Basically, RiskOptimizer incorporates probability data so the user gets not only a range of results, but also an assessment of the most likely outcomes. The ability to optimize using actual data from business records as well as most likely values of inputs while adjusting statistically for other possible values makes RiskOptimizer a better tool than its predecessors. Although this software requires considerable processing power, it is relatively easy to learn, and for many business problems, it is the most practical tool available (Seiter, 1999).

BroadVision's Enterprise Self-Service (ESS) Model is another business model software package that creates business value by transforming the way organization's do their business, that is, moving relationships to a personalized, self-service model that enhances growth, reduces costs and improves productivity. Leading global companies use BroadVision to power their Enterprise Self-Service initiatives — using the Web and wireless devices to unify and extend an enterprise's applications, information and business processes to serve its employees, partners and customers in a personalized and collaborative way. For example, BroadVision customer British Telecom (BT), concerned about call center costs, is converting key business processes to a self-service model. The net result is that significant cost reductions and increased revenues have been the result. BT's "Friends & Family" program costs an average of $4.73 per registration in the call center and just four cents on the Web. The BroadVision-powered BT.com Web site generates 100,000 new customer registrations each month and serves more than two million page requests per day (BroadVision, 2002). The above vendors, their products, and Web addresses are found in Table 1.

Optimization Software

In the development and implementation of optimal KM/WM systems, appropriate *optimization software* should be used when there is no easy, direct solution. This usually occurs when the problem structure is complex or there are millions of possible solutions. In such cases, there may be no efficient direct solution approach. Hence, optimization techniques can be used to search for the best solution. In some cases, when no solution can be found, the problem is relaxed, that is, some restrictions on the alternatives are lifted, and optimization is then used to find the best solution for the relaxed problem. Overall, optimization is the process of finding the best or optimal (near-optimal) solution from a set of

Table 1. Typical vendors, products, and Web addresses of business models and optimization software packages

Vendor	Product	Web Address
Aspen Technology Inc.	Aspen Marketplace Solution	www.aspentech.com
BroadVision	Enterprise Self-Service Model	www.broadvision.com
CAPS Logistics, Inc.	Supply Chain Designer Supply Chain Coordinator CAPS Logistics Toolkit	www.caps.com
Computer Associates	New Business Model	www.ca.com
International, Inc.	Erwin data modeling suite	
Ilog Inc.	ILOG Optimization Suite CPlex optimization suite OPL Studio 3.0	www.ilog.com
i2 Technologies	i2 TradeMatrix Plan	www.i2.com
Manugistics Group	Optimization modules	www.manugistics.com
Palisades	RiskOptimizer	www.palisade.com
Siebel Systems Inc.	Optimization modules, part of CRM (customer relationship management)	www.siebel.com
Trilogy	Optimization modules	www.trilogy.com

feasible solutions as set forth within the confines of the organization's business model.

Optimization software of all types is currently offered by a number of software vendors to solve a company's problems or assess a company's promising opportunities. For example, Aspen Technology Inc., CAPS Logistics, Inc., i2 Technologies, Ilog Inc., and Manugistics Group deliver optimization modules to the so called "process industries". Siebel Systems Inc., a leader in CRM (customer relationship management), has formed a partnership with Ilog that adds optimization capabilities to its configuration products. e-commerce enabler Trilogy Software Inc. keeps its framework open so it can work with a variety of third-party optimization libraries and routines. As with software for business models, these vendors along with their products and Web addresses are found in Table 1. Some of these vendors are covered below.

For Aspen Technology Inc., its software — The Aspen Marketplace Solution — combines best-in-class e-business technology and services from AspenTech and their partners to deliver next-generation private and public marketplaces supporting e-business collaboration among communities of process industry customers, vendors, and trading partners. AspenTech's solution is tailored to the specific needs and business processes of process companies. It provides a crucial link to AspenTech's supply chain and manufacturing optimization solutions, enabling process companies to respond to changing market demands faster and more profitably so that they can compete effectively in the new, networked economy by enhancing customer service and reducing operating costs. The Aspen Marketplace Solution incorporates AspenTech's process industry expertise, based on knowledge accumulated through its 20 years of experience delivering results-oriented solutions to the process industries. Creating value means addressing all three critical value chains — (1) supply chain, (2) engineering, and (3) manufacturing. To compete successfully and respond quickly in the new economy, process manufacturers must be e-business ready — automating, optimizing, and accelerating processes in all three-value chains throughout the enterprise and across the extended supply chain.

Supply Chain Designer from CAPS Logistics, Inc. (a Baan Company) is another optimization company. It is used by companies to make long-term decisions about how to place its manufacturing and distribution sites. Supply Chain Designer presents users with Windows-based wizards that walk the user through the basics of creating an optimization model. This is sufficient for typical problems, except for the more complex problems. Typical industrial solutions center on third-party logistics, consulting, consumer packaged goods, chemical, manufacturing, service, and transportation.

Ilog Inc., a leading French software house, markets its CPlex optimization engine. It has been used to assist in tasks that previously had been performed manually, such as package routing and production scheduling. Typical applications for Ilog optimization users include logistics resource allocation and scheduling, production planning and scheduling, manpower and service planning, and configuration and design. In addition, Ilog markets OPL Studio 3.0. Among other features, OPL includes a high-level modeling language that makes it easier to develop optimization models. Users have noted significant productivity enhancements after moving to OPL Studio 3.0, which centers on constraint programming. Another OPL feature is support for the Windows 2000 DNA architecture. With this feature, Ilog opens the optimization gates to Microsoft Component Object Model developers, Visual Basic developers, and, conceivably, Excel spreadsheet users.

Another software vendor i2 Technologies is offering a number of optimization packages. For example, i2 TradeMatrix Plan concurrently considers demand, materials, and capacities through the following business processes: (1) strategic planning, (2) demand planning, and (3) supply planning and scheduling. Strategic planning (first process) optimizes the supply chain network (locations and capacities of production and distribution facilities) to support the long-term business plan. i2 TradeMatrix Supply Chain Strategist compares the cost and service tradeoffs of network options. i2 TradeMatrix Supply Chain Strategist Simulator reports probabilistic characteristics of a network option with Monte Carlo simulations. i2 TradeMatrix Analyzer for Plan provides the capability to do advanced analytics and reporting on the plans and actuals for the Plan suite of products. Demand planning (second process) anticipates and influences demand. i2 TradeMatrix Demand Planner is a multidimensional forecasting environment that helps an enterprise to estimate and manage (through promotions and events) future demand. i2 TradeMatrix Merchandise Planner assists retailers in determining the most profitable assortment of products to carry at their stores. Supply planning (third process) determines what to ship, when, and where in order to meet projected demand profitably. i2 TradeMatrix Supply Chain Planner creates optimized master production, procurement, and transportation plans to meet anticipated demand and business objectives (Thierauf & Hoctor, 2003).

Goal Programming Software

A class of business problems that can be solved with the basic linear programming model by modifying the objective function is called *goal programming* (GP). Fundamentally, GP is a special approach to linear programming that is capable of handling a single objective with multiple subobjectives or multiple objectives with multiple subobjectives. Thus, higher-level objectives can be maximized or minimized first before lower-level objectives are brought into the final solution such that preference is given to those objectives that are of greater importance to an organization. These objectives are called *goals* because of instead of trying to reach the maximum or the minimum; a level of achievement (i.e., one or more goals) is to be obtained.

Although goal programming software, like linear programming software, has been applied to maximize a company's total operations for a forthcoming time period, it has been also successfully used in the selection of organization objectives that help determine what type(s) of objectives and strategies the manager should employ to realize the vast potential of an organization's resources. Other business areas where GP software has been applied are found in production-oriented problems and distribution ones. For example, goal programming has seen use in distribution problems where the manager has established a hierarchy of goals such as the achievement of a transportation budget, guarantee of delivery to specific customers, and utilization of certain routes due to prior agreements. The GP algorithm seeks the achievement of the goals in their order of priority; thus, higher priority goals are considered to be constraints that cannot be violated. Typical vendors that have included goal programming in their software are Baan, EDS, IBM, and SAP AG.

Product Lifecycle Management Software

Product lifecycle management (PLM) software was introduced in the 1980s and has been embraced by a number of organizations. Like ERP (enterprise resource planning), PLM software can be used by sales, manufacturing, finance, and service groups — any constituency that needs product information, including cost per unit and service history. The software presents a customized set of views and associated data to each type of users. One of the PLM's strongest selling points is enhanced collaboration. Users can check dimensions and engineering valuations in real-time with customers or among their own support engineers around the world. This compresses the introduction of new products and services by a considerable amount of time.

Vendors like UGS PLM Solutions (Plano, Texas), IBM Corporation (Armonk, New York), and SAP AG (Waldorf, Germany) are packaging PLM applications, business process workflows, and consulting services for specific vertical industries, like automotive, high-tech, and aerospace to provide a jump-start to manufacturers in those segments. Some, like Eigner (Waltham, Massachusetts), and SAP have targeted their PLM software to specific problems that customers are trying to solve. Eigner, for example, repackaged 40-plus PLM modules into eight products that cover everything from entry-level technical document management to maintenance, repair, and overhaul and on to configuration management. SAP recently announced the Collaboration Projects (eProjects) addition to its mySAP PLM suite, which functions specifically as a collaborative project management environment. Users want solutions to their problems so they can rollout PLM in a predictable, reasonable amount of time.

Going one step further, key enterprise systems (like ERP, CRM, and SCM) are becoming integrated with PLM software. A PLM package should be capable of talking to the CRM system for capturing customer's feedback, to the ERP system for capturing cost information, and to the supply-chain system for highlighting billing materials that can be reused or for advanced planning and scheduling. EMC, for example, was instrumental in helping build its Oracle Manufacturing Adapter, and other early adopters have paved the way for

Table 2. Typical vendors, products, and web addresses of product lifecycle management software packages

Vendor	Product	Web Address
Autodesk	Streamline	www.autodesk.com
Agile Software	Agile Anywhere	www.agile.com
Dessault Systemes	V5R11	www.3ds.com
EDS	EDS PLM solutions	www.eds.com
Eigner	Axalant	www.eigner.com
IBM Corporation	PLM Express, V5R11, SMART TEAM	www.ibm.com
PTC	Windchill PLM	www.ptc.com
SAP AG	MySAP PLM	www.sap.com

packaged connections to SAP, PeopleSoft, and a host of other enterprise applications. Partnerships like UGS PLM Solutions recently announced alliance with i2 Technologies (Dallas, Texas) and PTC's alliance with Siebel Systems Inc., (San Mateo, California) are also helping drive integration. In addition, most PLM vendors have announced relationships around middleware to ensure that their products link to legacy and other key systems (Stackpole, 2002). A listing of typical vendors, products, and Web addresses of product lifecycle management software packages is given in Table 2.

It is expected that PLM will expand to cover a much more comprehensive approach to an organization's operations. Fundamentally, because PLM is designed to gather, relate, appraise, and project all information, knowledge, and intelligence related to a product or a service over its lifecycle, this software needs to store many kinds of sales and cost factors pertinent to a product or service being analyzed. In turn, all of these facts are related to one another for an answer to the profitability or lack thereof in the years to come. Of paramount importance for accurate product lifecycle projections is information, knowledge, and intelligence about specific variables, assumptions, constraints, and like items to be considered in the modeling process of the product or service. Overall, a comprehensive PLM approach is called *venture analysis modeling* within an optimal KM/WM system environment. This approach will be stressed throughout the text.

Predictive Analytics Software

Another type of software found in optimal KM/WM systems is *predictive analytics*. Predictive analytics have successfully proliferated into applications that support customer recommendations, customer value and churn management, ad campaign optimization, and fraud protection. On the product side, success stories on demand planning, just-in-time inventory, and market basket optimization are a staple of predictive analytics. The real benefit from such analytics is to get to know the customer better and predict their behavior as well as forecast product demand and related market dynamics. Related to such analytics is the ability of companies to predict better the performance of their own operations that may depend on CRM (customer relationship management), SCM (supply chain management), and/or ERP (enterprise resource planning). It extends the reach of these systems by adding analytics that go into the fabric of a company's operations. For example, IBM has beefed up its core DB2 database engine with some of the analytical capabilities of Informix's database offerings as a result of its acquisition of Informix. Similarly, SAP has updated its business intelligence offerings, which include six new analytic applications for CRM, SCM, e-commerce, human resources, financials, and product lifecycle. PeopleSoft and Siebel Systems also have updated their offerings for predictive analytics.

Going one step further, software vendors are providing predictive analytics on the Web. At its core, Web analytics can be thought of as real-time support for decision makers. As an example, CRM lets marketing managers know who are their most profitable customers. In turn, they can use this information to up-sell and cross-sell their customers. Prior to the Web, CRM could be used to a limited degree. That is, marketing managers could track historical purchases, but could not capture buyers' reactions to products simply because it was impractical for a salesperson to follow a customer around a store. However, with Web analytics, it is relatively easy to tell which products its customers are looking at on the Web. By offering real-time information about how products and services are selling, the Web can foreshadow customer acceptance of new products and services. A clothing company, for example, can prelaunch a new clothing line online and use Web analytics to see how customers are responding. Online marketing campaigns can be refreshed every few hours, instead of waiting every few months. In terms of vendors in the analytics space, the traditional leaders — Accrue Software, Net-Genesis, and WebTrends — are being challenged by new players. Rather than focusing exclusively on Web analytics, many of these vendors are adding Web-usage data and common predictive analytic functions to their product offerings. The bottom line is that the Web-based predictive analytics software can dramatically improve the consistency and quality of decision making throughout an organization (Arena, Rhody, & Stavrionos, 2005).

Knowledge Discovery (Data Mining) Software

Basically, the underlying structure of data mining is, above all, *knowledge discovery*. The methodology of data mining involves the extraction of hidden predictive information and resulting knowledge from large databases. Because data mining is the process of analyzing data and information within a multidimensional framework, users can retrieve current detail data and information from within or outside the organization and summarize it by any desired category. Data-mining products provide a basic analytic capability (minimum, maximum, and average) and the ability to drill down to obtain more detail and to summarize details as necessary. Some data-mining products even go beyond basic analysis capabilities by providing statistical and mathematical routines to calculate the coefficients and powers of pre-specified independent variables. This is known as curve fitting or trend analysis. Trend analysis is used to determine patterns and relationships, and if key measurements are still within their limits and expectations.

When the variables are known, it is not difficult to determine patterns. Mathematical techniques are available to determine the relationship of one dependent variable (such as sales of a particular model) to several independent variables (such as the selling price, number of sales channels, inflation rate, etc.). However, the difficulty is identifying key patterns and trends when the analyst does not know the independent variables or, for that matter, may not even know what dependent variable should be analyzed. A data-mining product with knowledge-based capabilities can detect patterns in data and information that are not readily apparent, that is, it can indicate that there is a relationship between one dependent variable and one or more independent variables. Currently some data-mining products utilize data-visualization techniques to present these relationships graphically. The relationship need not be a causal one, nor is it most likely readily apparent. A typical knowledge-discovery tool is KnowledgeSEEKER, which can analyze and understand the patterns and relationships as well as be an accurate predictive tool. Numerous examples of KnowledgeSEEKER can be found in a text by one of the authors (Thierauf, 1999).

Today, data mining vendors are cooperating with computer vendors like Information Discovery Inc., NeoVista Solutions, Pilot Software, and SAS Institute, which are linked with Hewlett Packard. They are part of HP's OpenWarehouse Alliance Program, which focuses on selecting best-in-class warehouse component providers to ensure compatibility and to maximize functionality on HP computing platforms. For example, the Information Discovery System is a collection of client/server data-mining tools that work with Oracle, Sybase, and Red Brick relational databases and also perform data mining on the Internet. This software takes advantage of the parallel processing technology of an HP 9000 Enterprise Parallel Server by decomposing its analysis and sending it to the separate processors automatically. Also, the growth of e-commerce data generated by ERP (enterprise resource planning) and CRM (customer relationship management) systems require new data-mining technologies to help companies digest the wealth of customer behavior data. From this perspective, data mining takes on an expanded view of e-commerce data not found in previous information systems.

Data Visualization Software

One of the visualization tools useful in understanding simple to complex relationships within an optimal KM/WM system environment is *data visualization* software. Current data visualization is capable of representing integrated data intuitively for analysis, discovery, and research in databases, data mining, online analytical processing, and on the Web. Data visualization software has moved beyond simple charts and graphs that display a handful of data, that is, facts and figures. It has become the method of choice for understanding many complex business issues and activities. Pictures are no longer worth just a thousand words, but can be worth millions of dollars in the business value they can generate. To reap the benefits that are useful in developing and growing optimal KM/WM systems, decision makers can visualize multidimensional relationships through time that have always been there but never really seen. Various elements can be organized in a certain time frame or on a map to show travels over time, what was seen, who was met, what was done at first, what was done best, and how elements relate hierarchically. Facts can be organized to achieve new discoveries beyond the straight narrative of a story. Data has in it many valuable possibilities for creative manipulation in order to better understand the area under study (Brath & Peters, 2005).

Typical data visualization software includes the Cognos Visualizer (Cognos Corporation), IBM DX visualization library (IBM Corporation), ScatterPlot (Visible Decisions, Inc.), ADVIZOR Authoring Toolkit (ADVIZOR Solutions, Inc.), and Visual Mine *(AI* Software). For example, ADVIZOR Solutions, Inc. has assembled a tool suite/methodology called LISA (large integrated search and analysis) for tracking global trends and potential surprise events across various subject areas. LISA components have been orchestrated to support deeper interactions and integration to enable analysts to find desired information and knowledge, picture the relationships, realize the situation, foresee the outcome, model the possibilities, and decide a course of action. The picture component of LISA requires a visual interface that provides a wide array of visual displays. A single visual display is inadequate to solve and relate large quantities of unstructured data to the complex analysis process of anticipating the future.

Virtual Reality Software

Related to data visualization software is the world of virtual reality (VR). *Virtual reality* software is useful to create a walk-through scenario of how a company spent its money. Particular attention is given to flagging low-priority activities that were overfunded. Activities are assigned tall or short bars by their spending levels, and color-coded from green to red, to show priority. Exploring this 3-D landscape, managers are able to determine which activities to view, touch, and manipulate. For example, they can focus on tall red bars, indicated by low-priority activities with high costs. Touching those bars activates detailed information about the resources consumed. At any point, they could step back to see how resources were consumed for each high-level business process and for the entire company. Hence, virtual reality can be a very effective means to assist decision makers because it uses size, shape, and color in a 3-D landscape to convey useful knowledge that has been highlighted from extensive analysis.

Currently, designers of VR systems are moving more toward computer models that give superior depth perspective — even greater than that provided by the human eye. For example, Computer Associates International, Inc. has acquired two key players in 3-D media technology. An experienced creator of 3-D frameworks that model various real-world and imaginary environments is 3Name3D. Viewpoint DataLabs International, Inc. has developed a massive collection of the mathematical descriptions that capture the appearance and behavior of 3-D objects. Designers, instead of building physical models, are building them out of data, which gives them the flexibility to change things quickly and easily. DamilerChrysler, for example, uses this approach to reduce engine bay space and create larger passenger compartments. It also uses virtual reality to fine-tune its cars before they reach the clay model stages, and preliminary airflow and crash tests can be conducted without using cars or expensive wind tunnels and crash sleds (Thierauf, 1995). Going beyond these examples, virtual reality software has the potential to allow decision makers the ability to stand back and visualize ways of connecting points of wisdom not visualized before. The net result is that decision makers will be making more all encompassing and sound decisions for the areas under their control.

Overview of Effective Software for Optimal KM/WM Systems

Based on the above explanation of current computer software plus materials contained in the prior chapters, effective software for optimal KM/WM systems should provide decision makers with the ability to connect "points of wisdom" such that there is a balanced interaction among customers, suppliers, and the organization. Essentially this may mean starting with a new business model that takes a holistic approach to an organization's questions. The business model should provide for forging electronic partnerships with its customers and suppliers such that there is a win-win situation for all parties. If at all possible, these electronic relationships should provide an opportunity for the organization to collect on its accounts before payments are due. Sometimes, this is possible by the utilization of *optimization software* packages, which takes into account all facets of a company's operations as well as it numerous component parts. Where the parameters for the area being studied are somewhat difficult to measure, approximations should be made so that quantification is possible.

When utilizing new business models, it may be beneficial to include *goal-programming software*, that is, develop specific financial goals for the organization to achieve in the coming periods. This approach can be complimented by using *product lifecycle management software* that details the income, cost, and profit streams over the entire lifecycle of a company's products and services. In this manner, the details of the many products can be brought together for an overview of a company's operations. In turn, a company's decision makers can judge soundly, period by period, whether changes should be made to the multitude of products and services offered by the company. For more in-depth analysis, *predictive analytics* and *knowledge discovery software* can be undertaken to go at the company's underperforming products and services as well as those that outperform original estimates of sales, costs, and profits. In certain cases, it might be helpful for company decision makers to employ *data visualization* and *virtual reality software* to get a better handle on the company's operations that are good, bad, or indifferent. Appropriate software found in optimal KM/WM systems, then, are there to help decision makers judge soundly so that wise optimal decisions can be made today and tomorrow. As such, the software helps decision makers see the opportunity to be undertaken or the problem to be resolved by cutting through to the heart of the important relationships by connecting "points of wisdom."

Summary

Initially, the chapter discussed the importance of connecting "points of wisdom" in optimal KM/WM systems, followed by the changing nature of computer software today and tomorrow. Essentially the chapter focused on the utilization of a variety of computer software to build and grow such systems. These included the following software: new business models, optimization, goal programming, product lifecycle management, predictive analytics, knowledge discovery (data mining), data visualization, and virtual reality. Last, but not least, an overview of effective software found in optimal KM/WM systems was explored.

References

Alster, N. (2005, Summer). Software as a service. *CFO*, 41-45.

Arena, P., Rhody, S., & Stavrionos, M. (2005, April). Using advanced analytics to drive better business decisions. *DM Review*, 14-18.

Brath, R., & Peters, M. (2005, January). Information visulation for business: Past and future. *DM Review*, 40-43.

BroadVision. (2002, February 1). Enterprise self-service: A business model for a connected world. *CIO*, Advertising Supplement, S-1-S-4.

Computer Associates. (2001). How our new business model delivers value to customers, stockholders, and employees. *Computer Associates International, Inc.,* 1-13.

McAllister, N. (2004, December 6). Opening up the code. *InfoWorld,* 42-48.

Seiter, C. (1999, May). Complex business modeling made easy. *PC World*, 93.

Stackpole, B. (2002, August). What's the next step? Product lifecycle management. *Managing Automation*, 18-21.

Thierauf, R. J. (1995). *Virtual reality systems for business.* Westport, CT: Quorum Books.

Thierauf, R. J. (1999). *Knowledge management systems for business.* Westport, CT: Quorum Books.

Thierauf, R. J., & Hoctor, J. J. (2003). *Smart business systems for the optimized organization.* Westport, CT: Praeger.

Section III:

Building Optimal Knowledge Management/ Wisdom Management Systems

Chapter V

Development and Implementation of Optimal KM/WM Systems

Issues

- To examine the importance of the KISS principle in the development and implementation of optimal KM/WM systems
- To explore the need to take a holistic view to connect "points of wisdom" throughout the design process, to look at the cost justification of developing and implementing optimal KM/WM systems
- To set forth the detailed steps essential to develop and implement successful optimal KM/WM systems

Introduction

The design and implementation of a well-designed optimal KM/WM system requires taking an enlightened approach to the whole development process. That starts with applying creativity to the whole design process that is covered initially in this chapter, followed by the application of the KISS (keep it simple, straightforward) principle. In turn, EAI (Enterprise Application Integration) provides a framework for designing optimal KM/WM systems. Within this design framework, "points of wisdom" can be connected for producing optimal results. Next, the steps essential to developing and implementing effective optimal KM/WM systems are set forth. Included in these steps are those that relate to developing initial applications for connecting points of wisdom and disseminating results to grow wisdom of decision makers over time. Lastly, there is need to transform wisdom into action to meet decision maker's needs over time.

Creative Thinking Underlies the Development Process

Creative thinking, which was the focal point of Chapter II, discussed the importance of "innovation". The premise underlying creative thinking is that the innovation process is central to it and is too important to be left to chance. The innovation process is a rational approach to assist in developing a well-designed optimal KM/WM system. Such a system should connect points of wisdom that allows a company's decision makers to judge soundly about what needs to be done from the highest level down to the lowest level for growing an organization over time. Creative thinking approaches and their related techniques should be incorporated into an optimal KM/WM system so that optimal decisions can be made constantly by a company's decision makers. The complexity of today's constantly changing business environment and its impact on the organization should be an integral part of such approaches and techniques. By incorporating the required complexity in the design of an optimal KM/WM system, a company's decision makers will be better able to meet any challenges as well as surprises facing them. Additionally, a well-designed WMS has the effect of making a company's decision makers feel more a part of the organization. Incorporating creative thinking with new and different design approaches for making wise decisions by a company's decision makers is a must these days. In summary, creative thinking lies at the center of designing a system about what needs to be done. Wise, optimal decisions by decision makers drive the future growth of an organization.

Utilize the KISS Principle

Related to creative thinking is the KISS (keep it simple, straightforward) principle which is quite helpful in the design of an effective optimal KM/WM system. This principle ties-in with the concept that "less is more". Similarly, these two approaches to systems design can simplify the process of establishing relationships with customers, partners, and employees so that values desired by all parties are reached in an optimum manner. Research by Information Week reports that 70% of IT (information technology) and business managers surveyed agree that the company should be responsible for simplifying the process of establishing the partnerships. They placed less responsibility on the shoulders of industry consortiums or standards bodies. About half of the respondents say that either of those types of groups should take the lead in the simplification process (Schancupp, 2001).

Important Concepts Underlying the Development and Implementation of Optimal KM/WM Systems

Today, there are a number of important concepts underlying the development and implementation of well-designed optimal KM/WM systems. As a starting point, developers need to focus on an important concept found throughout the text, that is, adding values to all stakeholders such that there is a win-win situation. Essentially, the *value proposition* means that all individuals experience the benefits of a new product or service, that is, their life is made easier and less cost is incurred as a result of the new product or service. Tied in with the value proposition is the fact that business activities cross corporate boundaries these days, generating tremendous opportunities for using Internet technology to optimize business relationships. Overall, the value proposition establishes a framework for identifying the discrete, but interconnected activities that make a business operate and how those activities affect both the cost and the value delivered to buyers. The special advantage of the Internet and the World Wide Web is their ability to link one activity with others and make real-time activities created in one area widely available, both within the company and to customers, outside suppliers, and other channels.

Another important concept underlying the development and implementation of a well-designed optimal KM/WM system focuses on *"if it ain't broke, don't fix it"*. Too often, the prevailing thinking is that since the company is profitable, there is no need to make important changes to a company's basic business processes. This don't fix it attitude on the part or management needs to be replaced with a more proactive and positive attitude, that is, newer technology such as optimal KM/WM systems have the potential to improve the bottom line even more in the near future. Generally, a total transformation make sense more than a series of piecemeal adjustments or changes since the whole is greater than the sum of all its parts. The circumstances will dictate which approach is best for the organization to undertake.

An important concept underlying newer information systems, in particular, optimal KM/WM systems, is leveraging *open-source technology* to solve the enterprise technology problem in a very strategic and cost efficient fashion. Currently, the term open source means different things to different people. Information systems managers can no longer think in terms of a single vendor or hardware of a single operating system. They have to find ways to integrate an expanding assortment of computing products and services to meet the demands of enterprise-wide and even inter-enterprise information, knowledge, and intelligence in a competitive environment that has expanded to a global scale for optimizing a company's operations. This approach is found in an open-systems environment. Essentially, open-systems implement specifications for interfaces, services, and support in formats that enable properly engineered application software to be ported across systems, interoperate with other applications, and interact with users in a consistent manner.

In today's fast changing world, developers of well-designed optimal KM/WM systems need to be able to test and debug applications where they are going to be running, need to understand the business processes and implementation of those processes, and need a good understanding of the company's computing architecture. The architecture is very important because it encompasses so much more in a *Web-based environment*, especially for mission-critical applications. As a result, the analysis and design are generally very complicated in Web-based development due to the diversity and distributed nature of the players. Essentially, design becomes paramount because applications are built for the Web, thereby allowing customers to interact more closely with a company's business. Once developers have mastered analysis and design for Web-based applications, they need to master business-to-business applications that are found in newer information systems, especially optimal KM/WM systems. From this view, developers can take a holistic view of their business processes as they affect Web-based development that cuts across internal as well as external applications.

Related to a Web-based environment is the *real-time business* concept. Essentially, real-time business leverages a company's business connections through optimized business processes that can take advantage of instantaneous and near-instantaneous communication across all the components of a collaborative network and are supported by the relentless advances of new and more-focused technologies, products, and services. It is vastly different from the real-time computing concept of the past since a real-time business enhances and rounds out the business itself. To visualize a real-time business-operating mode, consider an auto manufacturer. From the customer to the salesperson to the sales manager who sends in the order for a specific model, to the order-entry clerk for a car manufacturer who uses the order to determine what is available from the multiple levels of inventories and on out to their suppliers. Instant, consistent, and, transaction-driven information links the ultimate supplier's supplier downstream through many levels to the final customer's customer. The bottom line is that all levels involved in providing a car to meet a specific customer order react in a real-time business-operating mode.

Another important concept that integration among software tools should be strong. One tool may require that data be imported and exported between tools while another offers bi-directional live links that are easy to use. If development tool vendors try to lock their users into their approach as a total solution, this might likely lead to problems down the road. Other factors that are germane to developing and implementing optimal KM/WM systems center on *scalability* or the ability to add hardware and software as system demand grows, *performance* in terms of meeting current real-time demands placed on the system, and *high availability* that includes online backup and recovery. Needless to say, other issues may need to be addressed — depending on the level of complexity before getting started on developing and implementing optimal KM/WM systems. Overall, critical decisions must be made about software tools, that is, wise decisions come from comparing alternative software tools and approaches after understanding each alternative. Complimentary to this discussion are the essential elements underlying developing and implementing well-designed optimal KM/WM systems, which are discussed below.

Enterprise Application Integration (EAI) Provides a Framework for Designing Optimal KM/WM Systems

Enterprise application integration (EAI) — as noted in Figure 1 — provides a framework for a holistic approach throughout the optimal KM/WM system design process. Since a comprehensive and holistic operating mode is the desired goal of optimal KM/WM systems, hardware and software products cannot be used in isolation, since building a corporate-wide enterprise application integration infrastructure requires the integration of many different methods, tools, techniques, and technologies. To unlock successfully EAI's collective expertise for company-wide usage, it is necessary to integrate a company's personnel and their related business processes with appropriate optimization techniques. There is need for company personnel to work in a collaborative environment. For an

Figure 1. Relationship of important factors that underlie an effective optimal KM/WM system design approach

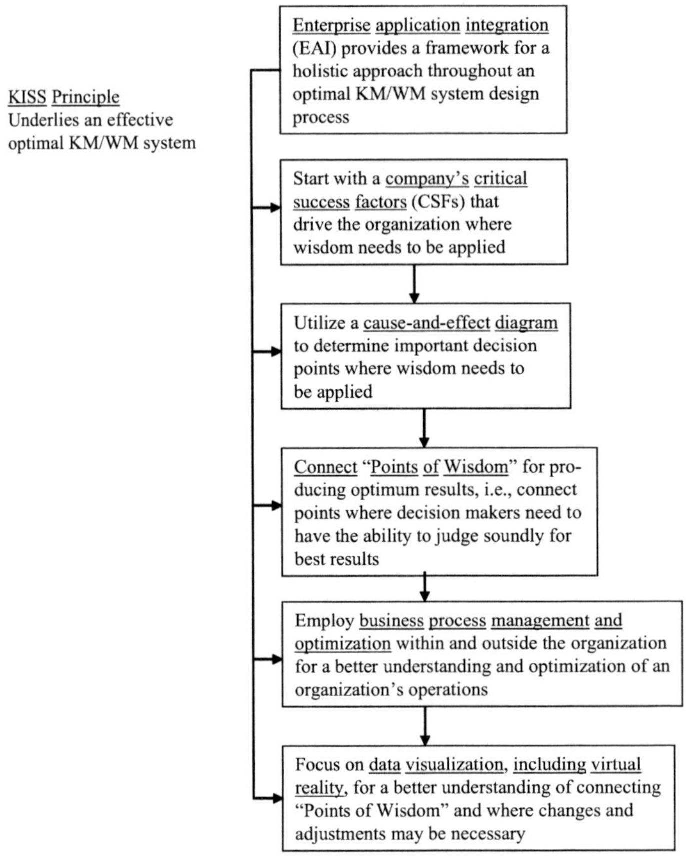

EAI environment to work effectively, it must be viewed by company personnel at all levels as a strategic means for the company to become more competitive to ensure the company's long-term success over time.

Because enterprise application integration is essentially the ability to read from and write to all the applications and sources across the enterprise, this integration supports unified views of information, knowledge, intelligence, and optimization results such that users can update facts and figures synchronously across systems. Although divisions can maximize their operations, decision makers can view information, knowledge, intelligence, and optimization results at a global level, assured that what is in one system is in sync with the rest of the enterprise's systems for optimizing a company's operations. With a commercial EAI product, there is a layer of software attached to each system that provides an interface from each application to an external integration system. Appropriate information, knowledge, and intelligence can be forwarded in a straightforward and simplified manner for making sound decisions about an organization's resources today and tomorrow.

Fundamentally, EAI is not just a set of software tools; it is instead a fundamental business process enabling the distribution and integration of the millions and billions of bytes of data generated throughout an enterprise. It builds a solid foundation — not only within a company's own infrastructure but extended across a company's supply chain as well — upon which decision makers can construct their business decisions with confidence. One of the benefits of EAI is that the integration is done in a fashion that allows systems to keep running even if another system crashes. The bottom line is that EAI is an ongoing process of building an infrastructure supporting internal and external operations, which can be created and changed as needed to meet the challenges of optimal KM/WM systems. Cutting across a company's boundaries includes Web services that improve service to customers. This approach gives a company unlimited reach and give business partners access to capabilities wherever they want to deploy them. In essence, this approach allows an organization to be where its customers and partners want it to be. The net result is that this customer and partnering approach provides for ease of management, cost savings, and productivity improvements for all involved.

It should be recognized that because technology can fail, it is important to have creative people on the design team. In essence, the design team should have the ability to improvise. The ability to improvise was dubbed *briocolage* by the French anthropologist Claude Levi-Strauss, but the concept could have just as easily originated from a CIO (chief information officer) watching its staff improvise their way out of obtuse technical jams on a daily basis. Many of the best solutions to technical problems have originated from *briocleurs* trying to solve a nagging problem with no obvious solution in hand. On the other hand, the issue is less of a technological problem than a financial one. Sometimes, the design team has to make do with what they have. The bottom line in either situation is using the collaborative creativity of the whole design team (Dickerson, 2002).

Typical EAI systems include a number of vendors currently. They are Active Software Inc.'s ActiveWorks Integrated System, IBM's WebSphere MQ Integrator, Mercator Software Inc.'s Mercator Enterprise Brokers, NEON's (New Era of Networks) e-Biz Integrator, STC's (Software Technologies Corporation) e.Gate Integrator, TIBCO Software Inc.'s BusinessWorks, and Vitria Technology Inc.'s BusinessWare. Overall, enterprise application integration makes a lot of sense if the organization is focused on simplifying and synchronizing a large number of applications within and outside the organization.

Start with a Company's Critical Success Factors that Provide an Overview of the Organization

Underlying an EAI framework is an appropriate business model that drives the organization. Related to this business model are *critical success factors* (CSFs) that provide an overview of the organization where wisdom needs to be applied (reference can be made to Figure 1). The CSFs are the important ones that are critical to the company's success today and tomorrow. Fundamentally, CSFs form the basis for giving decision makers the ability to judge soundly about an organization's operations. Generally, they give decision makers an overview of their organization and what really makes the organization successful. Among the important critical success factors today are:

- provide unique values to company stakeholders;
- maintain profitability that is at least equal to the industry's average;
- respond faster to competitive challenges;
- get products and services to the market faster;
- utilize a product lifecycle management approach;
- increase the quality of products and services;
- use information systems, such as optimal KM/WM systems, for strategic advantage
- increase customer loyalty and satisfaction over time;
- improve supply-chain operations with all parties; and
- provide higher productivity at lower costs.

As will be seen later in the text, CSFs are the real essence of decision making at the highest management levels which, in turn, drive the decisions at the middle and lower management levels of an organization.

Utilize a Cause-and-Effect Diagram to Determine Where Wisdom Needs to be Applied

To assist in locating where wisdom can be applied at specific decision points, it is helpful to take each critical success factor (they do vary from one organization to another) and draw a *cause-and-effect diagram* for each CSF. A cause-and-effect diagramming procedure (Figure 1) starts with listing all elements of one specific critical success factor. After each element is identified, it is numbered and circled, beginning with the number one. Next, arrows are drawn to show what causes what. For example, if element one causes element two, an arrow is drawn from the first circle to the second circle. When drawing the arrows, the question should be asked: "Which of the other elements cause or help cause this one?" After all arrows have been drawn, the *root* element becomes clear. It is represented by the circles with arrows leading away from it. In some cases, this cause-and-effect diagramming procedure can lead to other elements outside the initial inquiry.

To better understand this cause-and-effect diagram, reference can be made to the critical success factor noted above, namely, maintain profitability that is at least equal to the industry average. If a company is experiencing declining profits, causes can include the following:

- declining sales;
- poorly trained sales personnel;
- lower prices;
- increasing costs;
- inadequate inventory turnover;
- lack of product line for the times;
- increasing amount of older inventories;
- inadequate warehousing of finished goods;
- ineffective supply chain approach;
- too few sales personnel; and
- others.

Hence, a wide range of elements related to declining profits can be identified for review by management at all levels. Generally, declining profits can signal the need to review each area identified above in more detail to find the root cause or element. Hence, what starts out with the identification of one CSF, that is, declining profits, may lead to a series of related elements that need to be identified for the actual root cause. In this manner, important decision points need to be identified where wisdom needs to be applied.

Another example of utilizing a cause-and-effect diagram approach relates to the CSF — get products and services to the market faster. Underlying this critical success factor centers on determining what the customer wants before the individual is aware of a specific product or service. That is, a cause-and-effect diagram can be drawn that entails the effective and efficient coordination of the following major functions of an organization:

- corporate planning;
- research and development;
- marketing;
- engineering;
- finance;
- e-purchasing and supply chain;
- manufacturing;
- inventory;
- physical distribution;
- accounting; and
- human resources.

Although there is a logical flow from one major function to another, many of these functions, like physical distribution, can take many directions. That is, the products can be made to order, for stock, outsourced to other companies, or some combination. A cause-and-effect diagram can be drawn that depicts the most effective approach for operations today and tomorrow. Overall, a cause-and-effect diagram that is tied back to one or more critical success factors can be developed for determining the real root cause so that appropriate wisdom can be applied by a company's decision makers. It should be noted that a typical critical success factor and its cause-and-effect diagram along with connecting "points of wisdom" for producing optimal results (as noted in the next section of the chapter) will be set forth for Chapters VI through IX on corporate planning, marketing, finance, and manufacturing.

In addition, this cause-and-effect diagramming approach can be useful in identifying how information, knowledge, and intelligence is used and diffused, thereby helping direct the development of interfacing between the elements. When the elements are related to production, for example, a cause-and-effect diagram can provide a description of specific elements as well as a launch point for isolating problems in manufacturing operations. This can allow for tracking and isolating quality problems, mapping workflow and control back to information, knowledge, and intelligence usage, and addressing opportunities for optimization. Overall, this cause-and-effect diagramming approach provides a way to see how information, knowledge, intelligence, and optimization results can be used across the organization, identify gaps in their usage, and identify the processes involved in their integration.

Connect "Points of Wisdom" for Producing Optimal Results

Although the cause-and-effect diagramming approach does identify specific decision points where wisdom can be applied, there is need to get the organization's decision makers involved such that *"points of wisdom"* can be connected for producing optimal results (Figure 1). More specifically, these decision points represent the backbone of an optimal KM/WM system. They need to be connected such that decision makers can make sound decisions that take all of these connections into account simultaneously. In this way, the right hand knows what the left hand is doing so that decision makers can make wise organizational decisions.

In the area of marketing, for example, connecting points of wisdom go well beyond the confines of the firm and include a close relationship with customers, that is, anticipating their needs before they are aware of them. As a beginning point, a wise *market strategy* needs to be connected to daily *sales plans* which is backed up by appropriate *market research* and *pricing*. In turn, linkage to *advertising* and *selling* that is complimented by *sales order processing* and *customer service* is necessary within an optimal KM/WM system-operating mode. Connecting the foregoing major points of wisdom and making sound judgments about their operations is the best way to produce optimum results. Generally, these points of wisdom in marketing need to be related to those in corporate planning, manufacturing, and finance such that there is overall optimization of the organization's operations. Failure to include all aspects of a company's operations will generally lead to non-optimum or poor results.

In terms of physical distribution, for example, its major points of wisdom can be connected as shown in Figure 2. More specifically, the major points of wisdom that need to be connected for effective physical distribution are: (1) accuracy of sales forecasts on a build-to-stock basis and (2) finished goods exception reports, including back orders and out-of-stock conditions. The connect between these basic elements of physical distribution operations in Figure 2 centers on the right hand knowing what the left hand is doing. That is, accurate sales forecasts using a proven sales forecasting technique (such as from SAS or SPSS) should produce the desired amounts of finished goods at the right time and place. The numbers found in the finished goods exception report should be very minimal, that is, there should be very few deviations from what was forecasted. If this is not the case, there should be *real-time feedback* so that undesirable conditions regarding the sales forecasts are reported as they occur. Exception reports that include back orders and out-of-stock conditions should be reported as they happen. In this manner, physical distribution managers as well as their superiors

Figure 2. Connect "points of wisdom" for producing optimum results in the area of physical distribution on a build-to-stock basis

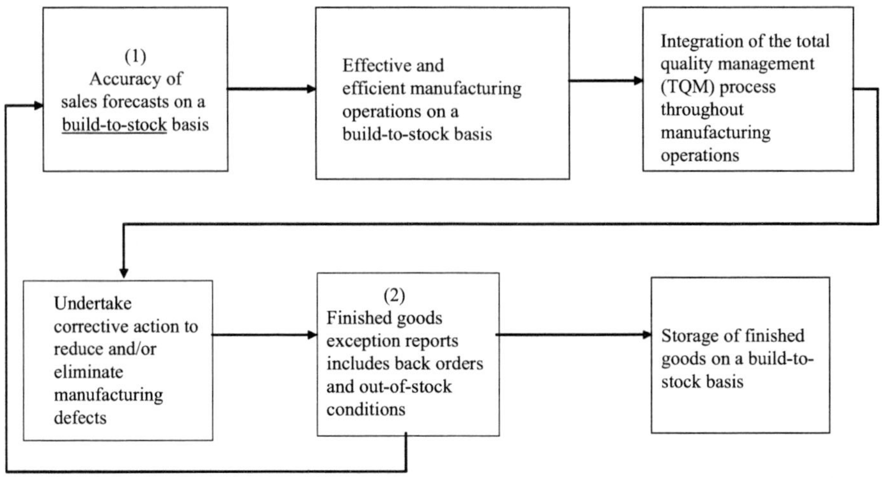

have the ability to judge soundly for best results as unfavorable conditions unfold in real time.

Employ Business Process Management and Optimization Within and Outside the Organization

Although companies create and implement strategies and business plans designed to optimize revenue, reduce costs, and provide desired profits, business-process management tools are needed to develop, integrate, and automate business processes that focus on connecting points of wisdom within the typical organization (Figure 1). Essentially, business processes can be defined as a set of interrelated tasks linked to activities that span functional boundaries. Business processes show how a company gets its work done — how it goes from taking an order to getting paid, or from selling a product to restocking its shelves. What is different about how companies approach business processes today is the way technology is woven into the efforts, thereby creating the opportunity and the means to reorient often long-standing practices. Tied-in with business processes are their management and the optimization on a continuing basis. *Business process management* (BPM) focuses on the ability of an organization to apply basic management principles (as set forth in Chapters VI through IX) to its daily operations. The orientation centers on optimizing overall operations that results in an optimum organizational efficiency and have a continuous improvement in quality. The end result is more-effective organizational operations.

The essentials of business process management, as set forth in Chapter III, center on the identification, comprehension, and management of business processes that interact with people and systems both within and outside the organization. They are based on the fact that businesses are ongoing, ever changing, and developing over time. Where one or more processes and their related points of wisdom ends, another begins. There is an understanding of, and catering to, process flow among multiple organizations and interested parties. A *business process management system* (BPMS) can be used to manage people-to-people, machine-to-machine, machine-to-people, and people-to-machine interactions. Basically, BPMS is the technology solution suite for a business process management approach to systems implementation. It is quite useful within an optimal KM/WM system environment since BPMS can be used to locate points of wisdom and change business processes so that decision makers can be assisted in making wise decisions. It should be noted that strategic KPIs (key performance indicators) and financial ratios can be developed for a BPMS to assist the development of a well-defined optimal KM/WM system (Griffin, 2004).

From another perspective, there is need to go one step further and talk about the optimization of business processes. Business processes can be in control, yet not optimized, that is, an organization can have well managed business-area activities, yet fail because the business-area activities are sub-optimizing the business areas' own work by not working as a team across the organization to optimize a process. Hence, there is a need to evaluate, minimize, and eliminate the activities that are adding cost. Creating duplicate databases is cost adding if data can be shared from a single or replicated database. From this view, *BPM optimization* fits within the framework of EAI. That is, the proper chaining together of discrete transactions, in the form of a business process, from one application to the next, constitutes enterprise application integration.

Currently, business process management products are available from major computer vendors. They include BEA Systems (BEA WebLogic Workshop), Collaxa Inc. (Collaxa 2), IBM Corporation (WebSphere), Microsoft Corporation (Microsoft BizTalk Server), and Sun Microsystems Inc. (Sun Web Service Choreography Interface (WSCI) Editor. Other BPM software vendors include Fuego, HandySoft, and Savvion Inc. (Ring, 2002).

Focus on Data Visualization, Including Virtual Reality, to Better Understand Operations

To help visualize and understand a company's business processes from an overview standpoint, *data visualization software* (as noted in Chapter IV) can be used to see relationships that have always existed, but never really seen (Figure 1). Essentially, various elements that relate to connecting points of wisdom can be organized in a relevant time frame or on a map to show travels over time which is seen, what is done, what is done best, and how elements relate hierarchically. These elements have in them the various possibilities for creative manipulation in order to better understand the proper time and place for connecting points of wisdom. A visual, three-dimensional look and feel using depth, height, width, texture, color, and so forth, allows users to view different aspects such that they have a different "landscape" for connecting points of wisdom. Overall, data visualization is a natural for a better understanding of the final optimal KM/WM system and its related elements.

Going one step further, virtual reality provides the capability to visualize complex data, information, knowledge, intelligence, and optimization results in an intuitive three-dimensional format. The interactive nature of the technology allows users to explore, with complete freedom of movement, either the complete world or specific areas of it. As systems become increasingly complex, virtual reality can assist in the complex system process. That is, it provides systems designers with real-time information in an easy to understand, graphical, and interactive format for a better understanding of complex business processes for connecting points of wisdom.

Use Application Service Providers (ASPs) as a Way of Expanding Wisdom

Application service providers (ASPs) have made their mark on the information technology landscape by allowing companies to leverage the technologies, processes, and expertise of the leading software platform providers — all without customers having to make the substantial investments usually required for such systems. Thus, the challenge is to identify, evaluate, and justify how a particular ASP meets an organization's requirements. Although application service providers have come under the name of outsourcing, in the past they have progressed from traditional co-located, hosted offerings to include client/server collaborative and enterprise solutions. In addition, there is a migration of the ASP market toward an application-neutral, distributed model on Web services. If such a model centers on one or more of the company's business processes or functions being considered mission-critical or if a large financial loss would result, it is wise to define a recovery strategy to minimize the company's risk. A standby arrangement with another service provider or an in-house product-based contingency solution must be considered for an enlightened approach to application service providers.

In addition, ASPs are shoring up global e-commerce applications. If a company wants an ASP to help shore up its global operations, there will have to be compromises. Delivering enterprise applications to multinational companies requires data centers and support teams in numerous countries, robust networks that span the globe, services that handle multiple languages and currencies, service levels that match needs in various time zones, and lockdown security. The benefit of moving to an ASP comes from off-loading the management of networks, computers, software, thereby letting the computer staff concentrate on more important tasks. The bottom line is the ability to turn an ASP's promise into reality that meets the needs of a company's decision makers, especially when considering its optimal KM/WM system's requirements.

Cost Justification for Optimal KM/WM Systems

An important question regarding the use of optimal KM/WM systems is, "How does one justify the cost in terms of a traditional cost-benefit analysis?" The answer given by them is that one does not. For the most part, the payoff from optimal KM/WM systems are different from other types of information systems in that they are often less tangible, less quantifiable. For example in the implementation of a new accounts receivable system, an information systems manager is able to estimate specific savings from tracking accounts receivable more closely. However, the new focus is on providing decision makers with the capability of judging more soundly their decisions for what needs to be done by employing newer techniques found in today's software that provide the best or near best operational results in the short- to long-run. The real payoff from an optimal KM/WM system is to give decision makers quick access to optimal results that lead to new ways in judging the soundness of their decisions as the learning organization grows over time.

It should be noted that there can be instances where savings or gains can be tied directly to optimal KM/WM systems. For the most part, however, no one has been able to come up with a model or system for cost justifying these systems. Because cost-benefit analysis is normally not used, the focus instead is on the value of these systems to decision makers, that is, managers, support staffs, and technical workers. Similarly, it is not easy to predict the total cost of such systems accurately. The largest expense generally comes from the on-going expansion of wisdom making decisions over time as decision makers become more familiar with the systems and seek new ways to judge the soundness of their decisions. Hence, decision makers have more relevant answers to opportunities and problems, that is, doing a better job of problem finding and problem solving. Typically, cost justifying optimal KM/WM systems tend to be evasive versus that found in traditional mission-critical information systems.

Steps Essential to Develop and Implement Successful Optimal KM/WM Systems

Currently, there are a number of suggested steps to develop and implement all types of optimal KM/WM systems successfully. Underlying all of these steps is the involvement of a company's customers, partners, and employees that gives them the ability to connect "points of wisdom" so that wise decisions can be made. The ultimate goal of an optimal KM/WM system for a typical decision maker is not only to do a better job of optimizing a company's operations so that it is run more effectively as well as efficiently in the short to long run, but also to impart wisdom so that decision makers are better able to judge soundly from a vast array of experiences acquired over time. The recommended steps that should be followed in developing and implementing successful optimal KM/WM systems are as follows:

1. Obtain support from the executive at the very top of the company
2. Designate a chief information officer to oversee the efforts
3. Appoint an experienced team to manage the development and implementation of the system
4. Utilize an appropriate methodology for connecting "points of wisdom"
5. Select an appropriate computer storage for developing wisdom over time
6. Utilize effective computer networking that assists in disseminating wisdom
7. Employ appropriate software packages that meet decision makers' needs for optimal decisions

8. Select high profile applications that connect "points of wisdom"
9. Distribute results to grow wisdom of decision makers over time

Each of these steps is covered next.

Obtain Support from the Executive at the Very Top of the Company

In the development and implementation of optimal KM/WM systems, the initial step is to obtain executive sponsorship, that is, to identify a *corporate sponsor*. No optimal KM/WM system will succeed without a strong advocate at the very highest level. It should be the number one or two executive in the organization. When top management is not there, the system is not going to go very far. Essentially, this is like trying to undertake development from a grass-roots perspective. Initially, getting started centers on obtaining the support of the president of a company or, at least, the support of the company's executive vice-president.

Designate a Chief Information Officer to Oversee the Efforts

Because sponsorship must go beyond a corporate executive sponsor and include a day-to-day operating sponsor, today this takes the form of a *chief information officer* who makes the initial request for optimal KM/WM systems and oversees their continuing development over time. Sponsorship means more than sitting back and saying I'm for this project and then delegating all of the work to others. A successful implementation requires a total commitment from the company's chief information officer who must equip the organization with the ability to respond to as-yet-unknown forces for change. Thus, the chief information officer must question assumptions, learn new technologies, and thrive in the face of uncertainty. The individual must be capable of operating comfortably in either the business or technological areas of an organization.

Appoint an Experienced Team to Manage the Development and Implementation of the System

Next, there is need to establish a project team that is capable of carrying out the implementation of optimal KM/WM system technology on a corporate-wide basis. The leader of the project team should be knowledgeable and receptive to newer computerized systems, have a good understanding of storage for aged and real-time data, have a good understanding of the latest in computer networking, and be receptive to applying newer sophisticated software packages. These areas were set forth earlier in Chapter III and IV. Ideally, the leader of the project team should have the time to work on the project full time, have an excellent understanding of both the organization's and management's decision-making needs, and have priority access to the chief information officer. To get an idea of the composition of the project team, a balanced mix of technology specialists and business analysts is needed. Technical specialists need to be recruited who are knowledgeable about new business models, optimization, goal programming, product lifecycle management, predictive analysis, business intelligence, data mining, data warehousing, computer networking, security issues, and other related areas. Similarly, business analysts who are knowledgeable about formulating objective functions and constraints for obtaining optimal or near-optimal solutions to a company's opportunities or problems need to be recruited in order to compliment the information systems professionals. Overall, the business analysts representing the various business/financial areas should be currently providing information, knowledge, intelligence, and optimization results to the company's managers that focus on the capability of decision makers to judge soundly on their decisions.

Utilize an Appropriate Methodology for Connecting "Points of Wisdom"

Related to the appropriate methodology to develop a successful optimal KM/WM system(s) is the consideration of whether to go with a "push" or "pull" philosophy for delivering optimal results to decision makers. If the push approach is utilized, this means that optimization of operations necessary for company employees, managers, or outside parties to do an effective job is relayed to their desktops. Essentially, optimization of operations is pushed forward to decision makers on a routine basis to assist them in their day-to-day tasks. If the *pull approach* is employed, decision makers are provided the tools needed to reach desired optimal results. In Chapter IV, software useful for an optimal KM/WM system-operating mode was set forth as possible candidates in this pull approach. Basically, two different system methodologies have a decided impact on the design architecture. The push or pull approach to total design can also be found in previous information systems that are upgradeable to accommodate optimal KM/WM systems.

In the design methodology set forth below, the push design approach is employed since the pull design approach is very open ended that varies from one installation to another. Part of this push approach is the development of corporate-wide storage of aged and real-time data. A well-designed, corporate-wide approach assists a company in such processes as data, information, and knowledge acquisition along with intelligence explanation, interpretation of optimization results, and validation of company strategies, procedures, rules, standards, and heuristics. Additionally, with the development of the corporate-wide data storage by the project team is the employment of user-friendly interfaces, that is, enterprise portals, as a means of getting to the information, knowledge, intelligence, and optimization results needed for optimizing a company's operations.

Using a push design approach, enterprise application integration provides an overall framework for designing optimal KM/WM systems. As noted previously an appropriate business model and optimization techniques drive the organization. In turn, a company's critical success factors provide an overview of the organization where wisdom needs to be applied. Based on a company's CSFs, a cause-and-effect diagramming approach can be developed from the highest level to the lower levels. In turn, appropriate "points of wisdom" can be connected by a company's decision makers in order to produce optimum results for a company's total operations. When making these connections, business process management and optimization within and outside the organization can be undertaken for best results. To better understand a company's operations, data visualization tools, including virtual reality, can be employed. Overall, the foregoing design methodology (as noted previously in the chapter) provides the means for a company's decision makers to make wise decisions that will have great impact on the organization over time. The preceding methodology must provide a basis for learning and the compounding of learning in order to produce the best optimal results based on wisdom for the current times.

Select an Appropriate Computer Storage for Developing Wisdom Over Time

To produce optimal results can occur throughout the organization, the project team that designs the computer storage should recognize that the short- to long-run needs of decision makers need to be considered before finalizing the content of aged and real-time data. To accomplish such a far ranging goal, there is need to consider e-commerce as it affects day-to-day storage. Today, accessing data warehouses for aged data and data federation systems for real-time data has been made a lot easier for a company's decision makers due to the pervasiveness of the Internet and the World Wide Web. The Internet has opened the door to decision makers where they are able to make better, far reaching decisions as well as move more quickly. Related to the Internet are intranets, which are private networks that leverage the Internet's infrastructure to extend corporate client/server networks to users wherever they may be.

Related to the data warehousing of aged data is the data federation approach, which is complimentary to enterprise application integration and the real-time business concept found throughout the text. A real-time computing system that does not use centralized query optimization and scheduling will retain local control while scaling to hundreds of machines. An enterprise-class data federation system supports dynamic load balancing across system resources. As loads on individual machines and networks change, the system adapts and adjusts query execution. As a result, the system can support many machines with high performance and throughput. From this view, this system is capable of obtaining the desired data in real-time or near to real-time as possible. Computer storage was covered in Chapter III.

Utilize Effective Computer Networking that Assists in Disseminating Wisdom

Designing optimal KM/WM systems means making minor to major changes to the company's computer networks. Typically, design changes need to tie-in with e-commerce so that a company and its customers can optimize their operations. Similarly, there is need for a company and its supply-chain partners to reap the benefits of optimization. E-commerce's success using this new design approach may, for example, lie in reaching untapped customers with innovative projects through community networking as well as taking advantage of the expectations that persistent connections will engender. Those customers who are not attracted by low prices and low margins are the high-margin buyers, expect quality products, high-involvement customer service, and personal attention. Generally, these types of customers require too much individual effort to be profitable in the current approach. Dealing with this kind of customer requires a new approach that delivers high or normal margins in exchange for real-time interaction, equal communication in both directions, the ability to learn enough about non-commodity items to risk buying them without touching them, and most of all, high-quality personalized customer service from a real person with expertise and authority whose functional intelligence is higher than that of a set of frequently asked questions. It results in a partnership between buyer and seller at a higher level that employs computer networking to better satisfy customer needs. Reference can be made to Chapter III for computer networking.

Employ Appropriate Software Packages that Meet Decision Makers' Needs for Optimal Decisions

The selection of the appropriate software packages centers on meeting the decision makers' needs for optimal decision making. The end result is an effective and efficient mechanism for producing optimal or near-optimal results. As such, decision makers ask the "why" of what needs to be done to develop opportunities and solve problems. In total, not only do these software packages need to be selected carefully to provide flexibility in optimizing a company's operations over time, but also this output should provide decision makers with sound decisions about what needs to be done for growing a learning organization over time.

Part of this discussion relates to the capability of real-time data analysis on the Web for analyzing a company's operations. As an example, the online viewer takes data input by users and produces charts, graphs, tables, and other graphic renderings. Many different features can be analyzed simultaneously and collapsed into subsets — for example, a year's buying patterns can be divided into quarterly tables. Users can sort numbers, drill down into charts, pivot dimensions, export data and print and create reports as they would with a desktop application. In, turn, these analyses can provide input for applying appropriate optimal KM/WM system software packages for optimizing a company's operations. It should be noted that software in the future might be bought and metered over the Internet as needed as electricity is done today. If so, appropriate software can be utilized within this operating mode (Garner, 2004). The subject matter of computer software for optimal decision making was covered in Chapter IV.

Select High Profile Applications that Connect "Points of Wisdom"

It is recommended that the project team start with high profile applications that connect "points of wisdom" for what needs to be done over time. Typical applications could start with a focus on a company's critical success factors (CSFs). Fundamentally, CSFs center on performing those factors that are critical to the survival of a company, such as customer satisfaction and product quality. An optimal KM/WM system approach could examine customer satisfaction with the view of optimizing their satisfaction as far as it is humanly possible. This may necessitate bringing the customer into the actual development process so that products and services are designed to meet their specific needs. In addition, this development process could utilize in-depth customer surveys, examine point-of-purchase information, utilize a product lifecycle management approach, form customer focus groups, and employ data mining or knowledge discovery techniques for connecting points of wisdom.

In a similar manner, this same type of approach could be applied to the area of product quality. Initially, the focus would be initially on customers in terms of how they perceive the quality of the company's offerings — good, bad, or indifferent. In turn, this could lead to quality improvements that impact what goods or services are really needed by a company's customers. Linkage to a company's customers is a must if the project team wants to show improved results for a company's initial high profile applications. In total, companies that utilize optimal KM/WM systems effectively should have the capability to add to sales and profits through greater use of getting the customer involved for added values that are perceived and recognized by them.

Distribute Results to Grow Wisdom of Decision Makers Over Time

The optimal results from high profile applications are based on their relevance to the times. Hence, appropriate utilization of final output from these applications is much more difficult than the simple capture of data, information, knowledge, intelligence, and optimization results. There must be procedures put in place so that decision makers can get at the optimized results for improving their judgmental capabilities over time. The distribution of optimized results means communicating the wisdom from vast experiences of not only "what" but also the "how" of what needs to be done. Typically, wise decisions come from a thorough understanding of alternative solutions and approaches to them. The distribution of output from these systems provides decision makers with the ability to expand their understanding of the company's operations over time as they discover new facts about their company that they did not think to ask in the first place. Thus, there can be new directions for transferring optimization results and, in turn, wisdom to decision makers that are not found in past information systems. Distribution of optimal results to grow the wisdom of decision makers is an important element in the development and implementation process of well-designed optimal KM/WM systems.

The Focus of Optimal KM/WM Systems is on Transforming Wisdom into Action to Meet Decision Makers' Needs

Related to the distribution of optimized results to improve wisdom of decision makers is the need to transform this wisdom into action for what needs to be done over time. The critical issue in placing decisions into action is finding the best balance between a payoff and cost. For example, a company has found that 30% of its customer base is the most profitable but it has a difficult time determining which 30%. An optimal KM/WM system-operating mode can help the company's decision makers find that segment and devise programs to maximize its sales and profits. Typically, the use of predictive analytics lets decision makers correlate customer transactional behavior in terms of how recently a customer purchased a product, how frequently a customer buys and the size of the transactions, to factors such as shareholder value, growth strategies, and other business concerns. Newer techniques embedded in appropriate software packages help decision makers discover new relationships, patterns, and trends that they generally would not see. Hence, it is necessary for the software to help decision makers distill wisdom that comes from a vast range of experiences over time. Even though a considerable amount of effort is required to get at the wisdom in order to make it actionable, the payoff can be extremely rewarding in that it lets decision makers keep their fingers on the pulse of the business for what needs to be done on a daily basis as well as over time.

Summary

The implementation of well-designed optimal KM/WM systems is quite useful to decision makers since it helps them judge their decisions more soundly than in the past producing optimal or near-optimal results. Initially, the chapter looked at creative thinking which underlies the development process and at the KISS principle needed for a well-designed system. Next, the important concepts underlying the development and implementation of optimal KM/WM systems were explored, followed by enterprise application integration which provides a framework for designing them. A focus on the big picture from a holistic approach throughout the design process was discussed. This included starting with a company's critical success factors, utilizing a cause-and-effect diagramming approach, connecting "points of wisdom" for producing optimum results, employing business process management and optimization within and outside the organization, and focusing on data visualization, including virtual reality, to better understand operations for making wise decisions. Also, cost justification was examined for these type systems. This background served as a basis for setting forth the essential steps in developing and implementing successful optimal KM/WM systems. As these systems develop further from a collaborative creativity viewpoint, there will be even more accent on a corporate-wide approach for judging the soundness of decisions that optimize a company's operations over time for what needs to be done.

References

Dickerson, C. (2002, September 9). Fostering creativity. *InfoWorld*, 56.

Garner, R. (2004, November 1). Software for hire. *CRN*, 19-22.

Griffin, J. (2004, October). Developing strategic KPIs for your BPM system. *DM Review*, 70.

Ring, K. (2002, September). What does business process management mean for the IT industry? *KMWorld*, 10-11.

Schancupp, P. (2001, April 2). There's no clear recipe for simplicity. *Information Week*, 114.

Section IV:

Applications of Optimal Knowledge Management/Wisdom Management Systems

Chapter VI

Optimal KM/WM Systems in Corporate Planning

Issues

- To contrast conventional wisdom with an enlarged view of connecting "points of wisdom" for corporate planning activities
- To examine those corporate planning areas that are critical to developing an effective corporate planning model
- To look at critical areas for short- to long-range corporate planning that are related to making optimal decisions
- To set forth a current optimal KM/WM system application in the area of corporate planning

Introduction

Leading futurists predict that the 21st Century will experience not only minor perturbations, but also major adjustments in business and social environments. Several of the driving forces behind these changes are global competition, the continual restructuring of business organizations, the aging of the world population, continued variations in the inflation (deflation) rate, the volatility of the stock markets around the world, globalization of capital markets, periodic energy shortages, and accelerating technological changes of all types. Within such an environment, future decisions will involve more complex ones than in the past and, to be effective, must merge together both quantitative and qualitative analyses. Developing new opportunities and solving problems for the typical company require the use of advanced computer systems, that is, optimal KM/WM systems, to provide top-level decision makers with a more effective approach to corporate planning over the short to long term.

Within this enlarged view of corporate planning, the chapter initially looks at the need to reinvent the organization for optimal decision making influenced by a number of management principles underlying corporate planning. The important elements necessary for the development of an effective corporate planning model within an optimal KM/WM system environment are set forth along with the model itself and its sub-models. Next, short- to long-range corporate planning is tied-in with executive visioning, problem finding, venture analysis modeling, and evaluating corporate performance. Finally, an optimal KM/WM system application that centers on corporate planning is illustrated for a holistic approach to an organization's operations.

Reinventing the Organization for Optimal Decision Making

To reinvent the typical organization for optimal decision making means starting at the corporate planning level. The continuing challenge for top-level decision makers is to anticipate, adapt to, and generate fresh ideas and approaches that exploit change for the times. The reinventing process may mean aligning a company's business with newer information technology, so as to be more agile with its customers and supporting partners. The bottom line is that companies need to stay relevant to the times about "what needs to be done" versus "staying efficient" which may not be the order of the day. In the information technology (IT) industry, for example, newer technologies will impact deeply the way people interact with computing devices of all kinds. Over the next several years, it is expected that the IT industry will embrace an optimal KM/WM system infrastructure, which will reinvent how companies create, sell products, and distribute its goods and services. Overall, a company's decision makers will need to reinvent their organizations for optimal decision making.

Conventional Wisdom Regarding Management Principles Underlying Corporate Planning

At the highest level of a company's operations, a number of management principles that are related to corporate planning have been developed over the years. Essentially, these principles are based on the utilization of conventional wisdom found in a typical company. These principles plus the ones presented in the next three chapters provide an underlying framework for a typical company whether it be manufacturing or service oriented. One of these principles which serves as a starting point recognizes that *change is a given*. From this view, there is a need to consider this factor in determining: (1) a company's mission and purpose (2) identifiable objectives, (3) measurable goals, and (4) strategies and programs. A company's mission and purpose gives overall direction to its entire organizational efforts.

For example, say that a company's mission and purpose is to be the dominant supplier of widgets to the most profitable domestic market segments and to provide a high-quality employee work environment. Identifiable objectives, regardless of how general they may be in their conception, become a measure of success in very real terms. They should be looked upon not as a reactive approach to setting direction for a company, but rather as a proactive force to realize organizational objectives successfully. These objectives, such as to compete in profitable market segments and be the low-cost producer in the company's market segments, provide the basis for determining measurable goals, such as increasing market share by 15% in high-profit market segments while reducing manufacturing costs by 10% per unit and achieving a 25% return on investment. In turn, these goals provide a basis for developing appropriate strategies that realize the company's mission and purpose. For example, strategies could include expanding market research activities to identify high-profit market segments and develop one or more product lines that fit the requirements and needs of these high-profit market segments. The successful linkage of a company's mission and purpose to its objectives, goals, strategies, and programs requires knowledge of a company's total business operations. The principle of effective utilization of a company's resources applies here and underlies the entire corporate planning function.

A principle underlying a company's long-term strategy centers on competitive forces. Competitive strategy is about being somewhat or completely different in the eyes of customers. Essentially, a company can outperform its rivals only if it can establish a difference that is perceived by its customers. This difference must also be preserved over the long haul. This approach contrasts with concentrating on operational effectiveness where techniques can be duplicated over time by competitors. Creating a successful long-term strategy involves choosing activities that are desired by customers and that are somewhat difficult for the competition to match. The essence of an effective long-term company strategy, then, is choosing to market goods and services differently from rivals. Otherwise, such a strategy is nothing more than a marketing ploy that will not withstand competition over time.

Related to the previous principle is one that allows a company to transform new technologies of products and services into an important advantage not only for itself but also for its customers. Too often, companies experience technological breakthroughs but are not capable of capitalizing on them. It cannot succeed without capable management to define or redefine the company's goals and its core competence. Without the proper managerial guidance, these pioneering efforts can result in disastrous forays into unknown territories. On the other hand, development of carefully prepared corporate short- to long-range plans that are supported by an appropriate organizational structure and effective marketing can thrust a company into a position of market leadership. Over time, it is necessary that technological pioneering efforts be maintained by continually bringing out important customer-oriented features ahead of competition.

For companies that are not engaged in technological pioneering but are operating in established industries, there is a need for top management to step back and take a hard look at its operations. This ties back to the concept of strategic thinking, that is, thinking in terms of strategies that are linked to today and tomorrow. Thinking involves identifying and mapping the driving forces that the company needs to evaluate. The mapping power provides top management and the corporate planning staff with visual representations of possible solutions and results, thereby helping them to achieve consensus in determining the driving forces. Once top management and its staff have formulated a new strategy, it must align the strategy with the company's resources to make sure that implementation is possible. In turn, this strategy can be translated into action by using aggregate project planning, as discussed below.

An important corporate planning principle is one that recognizes the need to develop capital project plans that focus on the development of new products and services. To map out and manage a group of strategic development projects effectively, there must be a total project plan in place. Getting to the point of an aggregate project plan is not easy, but working through the detailed process is a critical part of developing a sustainable development strategy. While the specific plan is extremely important, the planning process itself is even more so since the plan will change as events unfold and managers make adjustments. But choosing the right mix and determining the number of projects the resources can support, as well as picking the right projects, raise essential questions about how product and process development ought to be linked to the company's competitive opportunities. Overall, creating an aggregate capital project plan not only gives direction and clarity to the overall development effort, but also provides a framework for outstanding performance.

In the development of a company's corporate plans, the principle of examining the best- and worst-case scenarios comes into play. The best-case scenario is based on an economy that is experiencing a 3 to 5% growth rate where profits are very high. In contrast, the worst-case scenario considers the financial results of natural disasters like fires and floods. These are usually short-lived and can be solved quickly though at high cost. The type of disaster referred to here is conditions that prevent a company from fulfilling its long-term mission, objectives, and specific goals. Also, a worst-case scenario can refer to a reduction in business volume. Knowledge of these unfavorable conditions is necessary if an effective job of short- to long-range corporate planning is to be undertaken. While most corporate planners think in terms of good times, realistic projections should also include a provision for bad times, especially when global operations are used in developing realistic short- to long-range plans.

An integral part of the preceding principles is the element of risk. Fundamentally, a company's risk-management strategy needs to be integrated into its overall corporate strategy. A small company, for example, can ill afford to have "all of its eggs in one basket" since a disaster in the one area could bankrupt the company. In contrast, a large company can take on risks in one or more areas of its business. If this area(s) fails, it can be written off against other profitable areas of operations. However, a large company has to be cautious since the level of risk is affected by its competitor's strategies. Overall, an effective corporate planning principle states that the level of risk needs to be balanced with the company's ability to withstand significant losses and survive in the long run.

An Enlarged View of Connecting Corporate Planning "Points of Wisdom"

Typical management principles underlying conventional wisdom of corporate planning activities were set forth above. However, effective optimal KM/WM systems need to move to a higher level by connecting "points of wisdom" for assisting its corporate-planning decision makers in what needs to be done over time. A number of newer management principles underlying this enlarged point of view are found in Figure 1. First, an enlarged view of wisdom for corporate

Figure 1. Newer management principles help corporate planners get an enlarged view for connecting "points of wisdom" about what needs to be done over time

- A starting point for an enlarged view of wisdom centers on taking a comprehensive approach to an organization's corporate planning activities. What got the organization here today may not get the organization there tomorrow. Hence, there is need to connect "points of wisdom" for optimizing an organization's operations.
- A comprehensive understanding of a company's future direction helps a company to recognize what its customers want and need before they do. This understanding leads a company's top management and corporate planners to develop long-range strategic plans that provide a basis for action, In turn, these plans drive medium-range and short-range strategic plans.
- An enlarged view of wisdom cultivates a new style of management behavior that encourages changing established views of the company's products and services. New ideas and innovation can be the lifeblood of a company that takes in vast amounts of internal and external knowledge and intelligence to stay ahead today.
- Rethinking a company's operations include focusing on a venture analysis modeling (VAM) approach to a company's products and services. This approach should allow all interested parties the ability to come together for a win-win situation.
- Rethinking of a company's operations today is necessary in that the fastest growth segment of the U.S. population is 50 years and older, thereby displacing teens for the first time in the history of the United States.
- An important part of an enlarged view of wisdom in corporate planning is the need to invent the future that the organization wants. Even though this approach may appear to be risky, any company that sits on the sidelines will eventually realize the need for smartness and boldness of taking new directions for improving the bottom line.
- In the process of rethinking and possibly reinventing a company's operations, customers and their needs should be the real focus for change. A close relationship with customers can help the launch of a new product or service. By doing so, a company is connecting "points of wisdom" internally and externally.

planning activities centers an a comprehensive understanding of what it takes to develop and implement appropriate corporate objectives, goals, strategies, and programs over the short to long term. Optimal KM/WM systems should provide the ability to monitor business trends; to make intelligent strategic, tactical, and operational decisions based on uncertain judgments and contradictory facts; and to allow the organization to adapt quickly as situations change. Essentially, they need to rely on the exploration and analysis of related and unrelated information, knowledge, intelligence and optimization results to provide relevant insights, identify trends, and discover opportunities during normal times as well as during chaotic times and disruptive changes to an organization. As such, these newer systems center on serving the needs of the organization today for strategic corporate planning. In the process, there may be need to change the organization's business model for an improved strategic approach to corporate planning for tomorrow. What got the organization *here* today with an organization's business model may not get the organization *there* tomorrow needs to be reflected in the company's systems. Hence, there is need to connect newer points of wisdom for optimizing an organization's operations to their fullest.

An integral part of a comprehensive understanding of a company's future direction helps a company to recognize what its customers want and need before they do. (This concept will be explored in more depth in Chapter VII.) This proactive stance helps a company gain a competitive edge in the marketplace. At this point, it is safe to say that financial plans need to incorporate this important fact of competitive advantage. A company's short- to long-range corporate plans should also be related to its strategies so that the right hand knows what the left hand is doing. They should not only be realistic in financial terms, but also should challenge decision makers to meet financial targets that go beyond what a company has done in the past. Knowledge of past operations coupled with the challenge of the future should provide the company with opportunities to develop new ideas and concepts in order to meet these future projections. Essentially, long-range strategic plans provide a basis for action, which, in turn, drives medium-range and then short-range strategic plans.

An enlarged view of wisdom is one that cultivates a new style of corporate behavior that encourages changing established views of the company's products and services. There is also a need to encourage creativity in strategy and business processes. From this view, innovation is a style of corporate behavior that is comfortable with, even aggressive about, new ideas, change, and even failure. It must permeate an organization. Although it is not necessary to abandon a healthy business just for the sake of trying to be innovative, it is important to make sure that there are many good ideas in various stages of fermentation. When the time comes, the company needs to be willing to take risks. If a company waits to move until everything is certain, competition will have been there first. Thus, innovation can be the lifeblood of a company that uses vast amounts of internal and external knowledge and intelligence to keep ahead of competition.

As noted previously in the text, rethinking a company's operations includes focusing on a venture analysis modeling (VAM) approach to a company's products and services. In order for this to happen, a company's high-level decision makers must evaluate capital projects over their lifecycles for profitability before giving the green light for new products and services. Once the green light is given, the products and services must be evaluated constantly, that is, at least monthly and, in some cases, daily so that a company's product line is the proper one for the times. It should be noted that the competitive challenge for a typical organization in the future is more likely to come not from low-cost producers, but more from low-cost innovators — those that have the creative imagination to bring all interested parties together for a win-win situation in terms of profitability. The subject of venture analysis modeling will be explored in detail later in the chapter as well as in future chapters of the text.

Today, an enlarged view of wisdom, which requires a rethinking of a company's operations, must include the fact that the fastest-growing segment of the U.S. population is 50 and older, thereby displacing teens for the first time in the history of the United States. Needless to say, this shift signals very important changes for business organizations and the entire economy. For one, productivity for companies cannot possibly grow fast enough to offset the drop in the younger population and the increase in the number of older people. From this viewpoint, top-level decision makers within and outside the government should look to immigrants as a way out of this dilemma. In essence, without immigration, there can be no real growth.

A very important part of an enlarged view of wisdom in corporate planning is to *invent* the future that the organization wants. More specifically, if a company finds that its current products and services are not producing desired profit objectives, it is up to top-level decision makers to provide the wisdom and direction needed to reach desired profits with the company's talents and resources. Many times, this means rethinking everything. Newer technologies may provide the underlying structure to go off in new and profitable directions for a company. Corporate planners should pursue an aggressive, change-oriented approach to the development of new products and services that results in an improved bottom line for the organization. Even though this approach may appear to be risky, competitors who sit on the sidelines will eventually realize the smartness and boldness of new directions in paying big rewards.

In the process of rethinking and possibly reinventing the company's operations, customers and their needs should be the real focus for change. For example, technology companies have known that for years and tend to work closely with potential customers before any product launches. This is invaluable when a company is introducing a particularly complex product. In addition, a close relationship with customers can help in a successful launch of a new product or service. Hence, customers can be very helpful in assisting in a company's marketing efforts over time. The bottom line is that the organization is connecting points of wisdom internally with those externally in a more effective manner.

Development of an Effective Corporate Planning Model for Optimal KM/WM Systems

The development of an effective corporate planning model for optimal KM/WM systems must take into account the functional areas of an organization as well as how the model interfaces with external sources. The external sources of a typical organization include customers, government, public, competitors, suppliers, investors, financial institutions, and so forth. The major internal factors include marketing, finance, manufacturing, and human resources functional areas. In turn, these external sources and internal factors provide the needed input for developing an underlying structure for optimal KM/WM systems. In the discussion to follow, there will be a focus on the following: (1) make sense out of chaotic times and disruptive changes, (2) articulate a clear executive vision for the organization, (3) focus on executive leadership to execute this vision, and (4) employ corporate planning software useful in optimal KM/WM systems. Each of these areas is explored in some depth below.

Make Sense Out of Chaotic Times and Disruptive Changes

In order to make some sense out of today's chaotic times, such as the velocity and volatility with which trade, capital, and currencies move around the globe, decision makers need to utilize strategic intelligence to its fullest. As one approach, a company can intensify its intelligence gathering both internally and externally. In such a turbulent period, using published statistics to guide decisions is not the way to go. By the time employment figures rise or fall or retail sales rise or fall, it is too late. Typically, decision makers possess valuable untapped marketing intelligence within their own companies, if only they can get to it. The chief executive officer of DuPont, for example, holds a biweekly phone conference with 20 top managers around the globe to stay abreast of changes in customers, competitors, and local economies and politics. He asks different, pointed questions each time: What is happening to customers and their customers? What is the direction of local leaders to deal with the downturn? What should be done now to meet changing competitive rules? The sessions are not just for the CEO's enlightenment, but also for others to broaden their perspective of the global landscape.

Seizing new opportunities created by the crisis and, at the same time, stay relentlessly on strategy since all ships are being tossed by the same storm is another approach. The unsound may run into trouble and that can open up opportunities if a company is alert. At least three major financial services companies — AIG, Travelers, and Merrill Lynch — have acquired distribution systems in Japan as that country's finance industry has struggled. Cargill has been trying to crack the Japanese market for 30 years. Now the troubles of a Japanese competitor may finally give it a chance. Cargill announced recently that it would buy Toskoku, a Japanese food-trading company that has filed for bankruptcy. Tough times can also be ideal for forging new business relationships that will serve long-term goals that previously just had not seemed urgent. Current conditions have led Orion Capital, a Connecticut-based property and casualty insurer, to investigate strategic partnerships with several insurance carriers, distributors, and service firms.

Tightening up operations can be another effective approach, that is, there is no substitute for good business fundamentals — customer satisfaction, cost, quality, cycle time, and brand. Some ways of achieving them are smarter than others. For example, DuPont has centralized and streamlined its efforts to combine purchasing by diverse units around the world to get better prices from vendors — a process it calls "vendor convergence". Managers who see an opportunity to combine purchasing with another unit decide whether it would be the right move for all the businesses involved. They have ultimate authority, which they normally would not have to make this happen

Tied-in with making sense out of chaotic times is the area of disruptive changes in a typical company's markets. Disruptive changes can be caused not only by a number of factors that related to a company's markets, but also can related to a company's size per Clayton Christensen. When a company is young, its personnel, equipment, technologies, brands, and the like, define what it can and cannot do. As it becomes more mature, its abilities, for example, stem more from its processes, such as product development, improved manufacturing, and financial capabilities. Because companies, independent of the people within them, have capabilities, those capabilities also define disabilities. As a company grows, what it can and cannot do becomes more clearly defined in certain predictable ways. In the largest companies, values, particularly those that determine what are its acceptable gross margins and how big an opportunity has to be before it becomes interesting, define what the company can and cannot do. Because resources are more adaptable to change than processes or values, smaller companies tend to respond to major market shifts better than larger ones. Hence, it is suggested that companies capitalize on opportunities that normally do not fit in with their processes or values (Christensen, 1997).

Articulate a Clear Executive Vision for the Organization

Executive visioning is a most important element underlying corporate planning from the short range to the long range. An executive view entails farsightedness along with the eagerness to look ahead from a practical viewpoint. Effective executive visionaries are not necessarily those who can predict the 21st Century and beyond accurately. Rather, they are decision makers who can draw a conceptual road map from where the company is now to some imagined future, who can say, "This is how we get there." Visionaries come up with big-picture ideas that embrace the future, and they are able to get others to commit to the vision. As a result, employees feel connected to the company's future. As people with foresight and imagination, these leaders have a good sense of where the industry will be in the future and how a company should fit into that picture. They can show how a company's staff will benefit from the vision by clearly articulating the benefits. These people are charismatic and persuasive and have strong characters that inspire loyalty. All in all, visionaries have an agenda and use interpersonal networks to achieve their goals.

Essentially, an executive visionary needs to analyze his or her vision of the future by charting the company's future. That is, where does the company want to be next year, about three years from now, and five years from now, including a connection with the company's business partners? Can the company visionary say, "When people think of fulfilling their needs, they think of us first as the providers of quality products and services with reasonable prices?" If the answer is yes to this question, the company visionary is on the right track. Developing a vision means determining the image a company wishes to project to its customers, and then making sure that all of its products and services fit that image. Although the company visionary sees things in a different light from others, the individual is no mystic. The person's sources of information and knowledge are down to earth and extend beyond his or her gut-level feelings. Overall, a successful executive visionary must take in large amounts of information, knowledge, intelligence, and optimization results and not from inside himself or herself, but also from outside of oneself. In turn, this input must be digested which serves as a basis for judging soundly the need and the development of a company's future products and services alone or with the aid of business partners.

Focus on Executive Leadership to Execute this Vision

Vision by itself it not enough for the executive visionary to possess. The individual must be able to communicate what he or she has dreamed, and the company must have the required skills needed for its *execution*. The leaders of the organization must act consistently with the vision in everything they do. Too often in the past, top-management teams work up a statement of corporate vision, promulgate it, and then think their work is done. What they overlook, and what dooms this kind of superficial effort, is the need to plan and manage this vision over time. Newer computer systems, such as knowledge management systems, business intelligence systems, and smart business systems, are excellent vehicles for assisting in the fulfillment of carrying out the executive vision. However, more importantly, optimal KM/WM systems are designed to help decision makers to optimize the execution of their vision over time. In some cases, it may be necessary to reinvent the way a company markets its products and services for effective execution.

Going one step further at the highest level, execution has to be built into a company's mission and its purpose. Similarly, execution has to be built into a company's corporate objectives and goals, including its corporate strategies and programs. Execution should underlie a company's short- to long- range plans. In the authors' experience, companies that do not execute generally do not measure performance, do not reward employees, and do not promote people who know how to get things done. Typically, salary increases in terms of percentages are too close between top performers and those who are not. When a company strives for a culture of execution, there is a need to make it clear to all company personnel that rewards and respect are based on making sound judgment calls that connect points of wisdom. Overall, effective execution is required of all organizational decision makers to reach a desired executive vision.

Employ Corporate Planning Software Useful in Optimal KM/WM Systems

An integral part of corporate planning software is the whole concept of *corporate performance management* (CPM). Although all types of software can be used in developing and implementing the best-laid plans, there is still need to measure how successful or unsuccessful these plans have been. This is where *corporate performance management*, sometimes referred to as *enterprise performance management*, comes into play. A comprehensive CPM centers on evaluating what needs to be done for accomplishing desired results over time. Basically, corporate performance management extends across the entire value chain that drives a company's business. There is need to build relationships with customers that capture not only their spending habits, but also their loyalty over time. Additionally, understanding every link in a company's supply chain and evaluating those relationships not just to optimize a company's spending, but also to eliminate those suppliers that might put the company at risk. Overall, effective corporate performance management facilitates transparency of important information, knowledge, intelligence, and optimization results throughout the enterprise that are needed by corporate decision makers for judging soundly about what needs to be changed over time.

Going beyond the utilization of new business models and optimization techniques set forth in Chapter IV for a holistic approach to a company's operations, there is a wide range of corporate planning software that is available to top management and its corporate planning staff for developing and maintaining optimal KM/WM systems. Corporate planning software includes the following: (1) business planning, (2) goal programming, (3) product lifecycle management, (4) predictive analytics, and (5) balanced scorecard, all of which are covered below. Essentially, these software packages give company decision makers the big picture to assist them in judging the soundness of their corporate planning decisions. Some of these corporate planning software packages will be applied in this chapter.

A number of *business planning software* packages are currently available. Adaytum e.Planning 3.0, marketed by Adaytum Inc., is a financial planning software package that brings businesses closer to real-time financial planning. The software provides collaboration capabilities to connect more of a company's workforce into the planning process and integration features for cross-functional planning within companies. In the past, companies used slow manual methods for developing and communicating financial plans or business changes. The problem is that financial planners often do their work with little or no collaboration with sales, marketing, purchasing, manufacturing, and human-resource managers. Adaytum e.Planning 3.0 includes modules that let financial analysts, line managers, and executives participate in enterprise wide planning. The software combines top-down planning and financial modeling with bottom-up data collection, sales forecasting, budgeting, and workflow capabilities. That helps companies link strategic plans to operational initiatives. The software's e.PX extensions framework lets Adaytum develop modular, customized features for companies. The e.Planning application is linked to Adaytum's e.Planning Reporter or Microsoft Excel for real-time business intelligence. The system includes a data warehouse for storing planning data and XML Web-services support (Whiting, 2002).

Another business planning software package is Avantos ManagePro, which helps managers plan and track goals and progress while fostering success through focused management activities. It includes goal planning and tracking tools, such as the Top Level Goal Planner, for planning and delegating key business objectives, strategies, and tactics. The Goal and People Status Boards enable managers to monitor the status of primary business goals and obtain at-a-glance reinforcement of where to focus attention. Still another business planning software package is Jian BizPlan Builder which is a complete business plan template on disk with more than 90 typed pages of example text that are formulated into word-processing files. Templates use standard spreadsheet applications like Excel or Lotus 1-2-3 to calculate financials and generate graphs. The step-by-step format guides the user through, explains issues, and gives clear and sensible advice.

Once a company's business plans have been developed (from the short range to the long range), software can be employed that is useful to determine the company's overall financial performance. Such software goes beyond monthly actual versus budget reports that are routinely produced by today's information systems. The balanced scorecard — as discussed below — has quickly gained wide acceptance in many companies around the world, independently of their size and business sector. Since a fundamental application of the balanced scorecard is to communicate strategic objectives to employees at all levels in an organization, there is a special opportunity to introduce systems dynamics as a common language known by everybody as flowcharts. As company employees gain a better understanding of the whole organization's dynamics, they will be able to find new opportunities for contributing to enterprise objectives (Kaplan & Norton, 2001).

Goal programming (GP), which was noted in Chapter IV, can be quite helpful to top-level decision makers in the area of corporate planning. Because GP is capable of handling a single objective with multiple subobjectives or multiple objectives with multiple subobjectives, higher-level objectives can be maximized or minimized first before lower-level objectives are brought into the final solution. From this perspective, preference can be given to those objectives that are of greater importance to an organization from a corporate planning standpoint. As such, a company's objectives and measurable goals can be connected mathematically so that a company's profitability can be pursued in an optimum manner over the short term to the long term. Goal programming, then, is useful to decision makers in helping them to determine what type(s) of objectives and strategies the decision maker should employ to realize the vast potential of an organization's resources. Similarly, sales goals or quotas can assist decision makers in determining the time to be spent on established customers versus new ones.

Currently, several major software vendors (mentioned in Chapter IV) have included goal programming as the central focus of their generalized problem solving techniques. Goal programming has been used to integrate sales goals with production goals such that both types of goals can be realized to the fullest extent possible. An example of this approach will be given in Chapter VIII. In addition, goal programming has been included in multiple criteria decision making techniques as a way of maximizing or minimizing goals. Integer goal programming has been used to determine the best combination of multiple products to meet the demand for items and materials with a variety of physical, financial, and environmental goals and constraints. In a similar manner, fuzzy goal programming, which is an extension of integer GP, introduces vagueness into the modeling process. Overall, current approaches to goal programming involve solving problems as well as exploring opportunities involving the realization of multiple objectives that operate within a number of constraints.

In terms of a venture analysis modeling approach, its appearance is being widely employed by a number of organizations in the form of *product lifecycle management* (PLM) *software*. This type of software was introduced in Chapter IV. A PLM approach is not a complete shrink-wrapped solution currently. But rather, it is a conceptual view of a manufacturer's product addressing and incorporating the strategic business objectives of the company. PLM creates a framework for collaboration and sharing of product information among strategic partners, both within and external to the enterprise, over the course of a product's life — from initial concept to retirement. Essentially, it is business-driven and technology-enabled.

Current PLM software is available from a number of vendors. Two representative vendors are set forth below. PLM for Baan is a process-based enterprise strategy that leverages a foundation of existing tools, processes, and knowledge assets to help an organization achieve better efficiency, productivity, competitiveness, and higher margins. This vendor's approach to PLM includes document and product data management solutions, a workflow engine, real-time analysis and reporting tools, and *global collaboration* capabilities. It also includes a wide range of out-of-the-box integrations to ERP, CAD, Office, CRM, reviewing and redlining systems and others, thereby allowing controlled, predictable, and progressive projects with a high return.

mySAP Product Lifecycle Management (mySAP PLM) is another software package that is a business solution, which enables collaborative engineering, custom product development, and project, asset, and quality management among multiple business partners. mySAP PLM integrates all participants in the development process: designers, suppliers, manufacturers, and customers. Engineering is no longer a linear value chain, but a three-dimensional, collaborative community focused on a common goal. This software provides an ideal solution for any company that requires proactive quality management to improve equipment and plant performance. And the solution is ideal for any industry in which maintenance costs and equipment reliability directly affect profitability. mySAP PLM is delivered through mySAP Enterprise Portal — a role-based portal solution. Users get convenient, Web-based access to the internal and external content, applications, and services they need to do their jobs. It is integrated with mySAP Supply Chain Management, mySAP Customer Relationship Management, mySAP Supplier Relationship Management, mySAP Exchanges, and SAP's enterprise resource planning (ERP) offering to support collaborative engineering, design, and procurement (Stackpole, 2003).

Ideally, a broad-based product lifecycle management approach is able to gather, relate, appraise, and project all types of information, knowledge, intelligence, and optimization results associated with a product or service over its lifecycle for optimizing profits. As such, the software stores many kinds of sales and cost factors pertinent to a product or service. Within a comprehensive PLM approach, decision makers can make solid connections between sales and costs in years to come for its products and services, that is, connect points that determine where future profits will come from. In essence, a broad-based product lifecycle management approach, that is, a venture analysis modeling approach, may assist company decision makers in connecting points of wisdom for today and tomorrow.

Predictive analytics software is useful to a company's decision makers in a number of ways. It can be employed to assist in determining the sales, costs, and profits of a product lifecycle management approach — as discussed earlier. Or this type of software can help a company's top-level managers determine which of their customers are most likely to respond to a sales pitch, switch to a competitor, or prove to be unprofitable through nonpayments and defaults. Basically, analytic tools use data mining or knowledge discovery to build models that examine historical data such as customer's purchases and use that information to predict future behavior. Current suppliers of predictive analytics software include Fair Isaac, Genalytics, IBM, Oracle, SAS Institute, and SPSS. Genalytics current 5.0 version has proved to be helpful for financial services, telecommunications, and retail organizations. SPSS Predictive Marketing 2.0 is designed to help companies with multiple marketing programs determine which should be directed at individual customers (currently a complex labor-intensive process).

Corporate *balanced scorecard* (BSC) *software* was first introduced in the early 1990s. Essentially, the BSC approach has evolved from an enterprise-wide performance measurement system to a *corporate performance management* (CPM) system. Used to define targets and metrics that help align resources with goals, the BSC is increasingly being applied to the management of those customer-based business strategies as well. The reason is that when managed across the enterprise, customer-based initiatives are easier to measure, justify, and secure funding. In addition, the rule of technology in supporting a customer strategy is better understood internally. Constructing a balanced scorecard approach means bringing its four components — finances, customers, business processes, and learning factors — into alignment. Financial results can be delivered by supporting the customer-management process and transforming the customer value proposition. Each of the four propositions brings to the surface a question that must be answered in the course of selling the strategy. The enterprise-wide management of finances, customers, business processes, and learning factors can improve a company's execution of corporate strategies and related programs (Norton & Russell, 2003).

Essentially, a BSC approach supplements traditional financial measures by measuring the important items that matter to a company and its clients. The balanced scorecard enables a company to track financial results while monitoring progress in building the capabilities they would need for growth. A BSC approach can be looked upon as a control panel of an airplane by keeping track of a company's financial progress as well as its softer measurements. The bottom line is that a balanced scorecard approach can integrate corporate strategy with the wisdom of managerial decision makers for what needs to be accomplished over time. Current BSC vendors include CorVu, Gentia, Hyperion Solutions, Oracle, PeopleSoft, SAP AG, and SAS.

Corporate Planning Model and Its Sub-Models Found in Optimal KM/WM Systems

In the development of a corporate-planning model and its sub-models as found in optimal KM/WM systems today, a logical starting point is the need to add unique values for customers that tie-in with external sources and internal sources such that there is a win-win situation for all stakeholders. This is the approach taken in Figure 2 for connecting "points of wisdom" for corporate- level decision makers within an optimal KM/WM system environment. Later in the chapter, connecting corporate strategy to points of wisdom for better optimization of a company's resources will be set forth. Before these points of wisdom can be connected, there is a need to start with executive visioning under the direction of top management and the corporate planning staff. The *executive visioning sub-model* sets the overall direction for the organization today and tomorrow. In some cases, executive visioning may call for a new organization business model. In other cases, there may be need for revisions to the business model. Or, it is possible that the present organization business model is proper for the times. It should be noted that turbulent times call for some type of change to the organization's business model.

The next sub-model shown in Figure 2 is the *corporate mission and purpose plus corporate objectives and goals sub-model*. A company's *corporate mission* statement provides the operational substance for a company, that is, who the customers are, what customer needs are being met by the company, and how the company can fulfill those needs. Related to a company's mission is its purpose for being in existence, that is, what useful need it provides for its customers. *Corporate objectives* serve as guidelines for action. More importantly, they serve as a standard for measuring performance and can provide

Figure 2. Relationship of corporate planning to a company's functional areas for connecting "points of wisdom" within an optimal KM/WM system environment

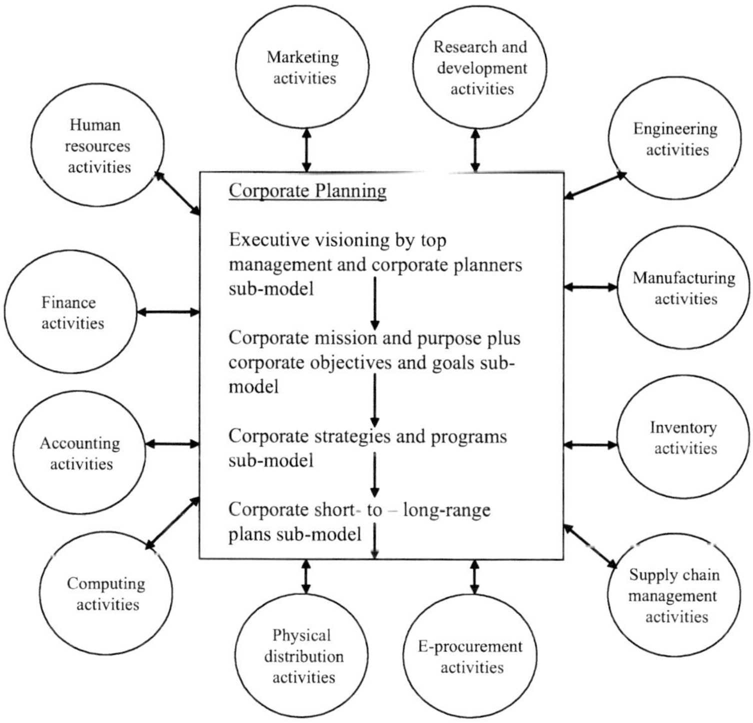

incentives for employee performance. As such, most objectives take the form of corporate goals that are attainable, quantifiable, and measurable. Furthermore, every business needs strategic objectives and goals as well as financial ones. A strategic orientation might include a larger market share, quicker design-to-market times, superior customer service, and the like while a financial orientation might include revenue growth, earnings growth, and stronger cash flows.

Strategic and financial objectives and goals, in turn, can be set forth as specific corporate strategies and programs that are needed to allocate a company's resources in an optimum manner. This is the content of the *corporate strategies and programs sub model*. Typically, *corporate strategies* emerge over time as an organization innovates and responds to its markets. Effective strategies develop through experimentation, through design, or a combination of the two. The best strategists see patterns taking shape in their environments. In this way, they find strategies created over time that can be used to determine where corporate funds need to be spent on corporate programs to optimize a company's operations. Allocation of a company's resources via its programs provides a basis for short- to long-range financial plans. Continuing per Figure 2, the *corporate short- to long-range plans sub-model* centers on the organization's financial plans today and tomorrow. More to the point, projected production, manufacturing facilities, and personnel requirements in the medium- to long-range are determined along with projected balance sheet and income statement plus source and application of funds. Additionally, *short-range financial plans*, that is, current year, include monthly balance sheet and income statement plus product lifecycle management reports, cash flows, and pertinent exception reports. Essentially, management needs cost and revenue margin information and knowledge so that it can identify areas of strength and weakness. Feedback from these short-range sources of information and knowledge provides a means for determining whether or not a company's management is doing its job as well as whether or not changes need to be made at the highest level on down to the lowest level. It should be noted that corporate policies and procedures are needed to carry out corporate short- to long-range plans. Where corporate policies and procedures are not in conformance with those set forth by company decision makers, evaluation could lead to changes essential to optimizing a company's operations.

Short- to Long-Range Corporate Planning

Because short- to long-range corporate planning is at the center of an organization's operations, it is of importance within an optimal KM/WM system environment since points of wisdom need to be connected for a better optimization of a company's resources. Corporate planners need to generate fresh ideas, which can be translated into new opportunities that challenge the organization. As such, appropriate corporate planning software, like goal programming and product lifecycle management, can be applied. Underlying corporate planning is setting or changing organization objectives and goals as deemed appropriate that are based on a company's mission and purpose, obtaining the resources to meet these objectives and goals, and determining the corporate strategies and programs that are integrated into corporate short- to long-range plans. Since corporate planning occurs at the highest level and is related directly to top-level executives and their corporate planning staff, the appropriate external sources and internal factors are merged that are critical to setting the proper present and future direction for the organization. These important elements of corporate planning are incorporated in the following exposition that centers on judging the soundness of decisions at the highest level for helping the organization better optimize its resources.

Tie-In of Executive Visioning with Corporate Planning for the "Big Picture"

Getting the "big picture" is an essential element of executive visioning at the corporate planning level. As will be seen in future chapters, this means going beyond what is happening currently and expected to happen in the future, but also what had happened in the past. An enlarged view of wisdom recognizes that history has a tendency to repeat itself. As Peter Drucker noted, the greatest weakness of American business is that it knows no history. Although the repeat will not be exactly alike, there will be parallels that will be obvious to the intelligent viewer. For example, there are some telling parallels between the current Information Technology Revolution and the Industrial Revolution of the 19th Century (Butts, 2002). Overall, getting the big picture is essential before executive visioning can take place. Related to executive visioning is problem finding — covered in the next section.

An important element underlying corporate planning from the short range to the long range is *executive visioning* (noted previously in the text). An executive view entails farsightedness along with an eagerness to look ahead, but, at the same time, it must be practical. Effective executive visionaries are not necessarily those who can predict this century accurately. Rather, they are the decision makers who can draw a conceptual road map from where the company is now to some imagined future, who can say, "This is what needs to be done and how we get there." Visioning implies a change from the status quo, which helps explain why visionaries are over represented in the ranks of entrepreneurs. It also explains why they come in handy to an organization in deep trouble — think of Mr. Lee Iacocca saving Chrysler in the last century. Vision is not for the complacent. While the executive visionary sees things a bit differently, this individual is no mystic. The person's sources of information and knowledge are down to earth customers and suppliers, for example, and extend beyond his or her gut-level feelings.

Executives with vision typically share a couple of other characteristics. They have a high degree of self-confidence. It takes a lot of inner strength to imagine a future at variance with the common expectation and to sustain that vision in the face of responses ranging from incredulity to derision. The visionary may also be a bit of a loner. Ego strength can mean that the individual has less need of other people. But executive vision has a way of exciting others somehow. An executive visionary appeals to the emotions and aspirations of people in a way that goes beyond the usual carrot and stick approach.

Most corporate visions start out as desirable and obtainable strategies. William Hewlett and David Packard established a company that is dedicated to producing the best instruments for the world's scientists. That surely counts. Ray Kroc built McDonald's and Kemmons Wilson founded the Holiday Inns chain — both visionary achievements of American capitalism. Vision by itself, however, is not enough. The executive visionary must be able to communicate what he or she has dreamed, and the company must have the required skills to execute the dream. The leaders of the organization must act consistently with the vision in everything they do. Too often in the past, top-management teams have worked up a statement of corporate vision, promulgated it, and then have thought their work is done. What they overlooked, and what dooms this kind of superficial effort, is the need to plan and manage this vision over time. An optimal KM/WM system is an excellent vehicle for assisting in the fulfillment of the executive vision.

It should be noted that executive visioning currently includes all personnel who work for multinational firms that have a global mind-set. It is not enough that a few top-level decision makers have a global mind-set for a company's wide-spread operations. All employees should excel at balancing global consistency with local responsiveness. That is why many organizations are testing ways to embed a corporate global mind-set in company-wide policies. Research shows that managers universally recognize the need, but only a few are close to implementing it. To assist company employees in implementing a corporate global mind-set, an optimal KM/WM system can be an excellent vehicle for doing so (Begley & Boyd, 2003).

Utilize Problem Finding to Turn Problems into Important Opportunities

For problem finding to be truly effective tomorrow, corporate planning must be undertaken from a *proactive* standpoint. External sources and internal factors that affect the organization, from the short range to the long range, are also taken into consideration. Many times, such information, knowledge, intelligence, and optimization results are found in the company's historical data warehouses and real-time databases. The most important external sources focus on customers, suppliers, the government, investors, the public, financial institutions, and competition. The most important internal factors center on organizational strengths and weaknesses, objectives and goals, functional areas and their personnel. Obtaining and using this information, knowledge, intelligence, and optimization results are an integral part of the corporate planning process. Overall, the raw material of corporate planning is needed by top management and its corporate planning staff to initiate problem finding. Likewise, it is the means by which decisions at all levels of the organization are made to further the organization's objectives and goals.

To have an effective planning process, top-level decision makers and corporate planners need *to identify potential problems* in the future and bring them back to the present for resolution. Problem finding tends to center on finding solutions to problems that many deem impossible to solve. Going one step further, there is the need *to identify opportunities* that are related to future problems. As in the past, top-level decision makers and corporate planners must also be involved in problem solving. Thus, they need to go beyond problem solving and be actively involved in problem finding to allocate and use the organization's resources effectively.

To assist this highest-level group in problem finding, predictive analytical tools as well as mathematical and statistical models can be employed to develop appropriate analyses. Because the highest-level problems tend to be semi-structured and unstructured, a series of "what if" questions can be asked and answered to get a feeling for future problems that are poorly defined. Hence, top-level decision makers, when they are oriented toward the near future, need to answer pressing questions. However, as the time frame is extended further into the future, the problems become even more unstructured and may require heuristics (i.e., rules of thumb) based on present and newer corporate planning techniques. Due to the vast number of uncertainties over time, the reliability of answers to these "what if" questions are, needless to say, less accurate. Nevertheless, top-level executive needs are being met as well as possible based on the magnitude of the unknowns. For more information on problem finding, reference can be made to Chapter II.

An important benefit from problem finding is that a manager can prevent a molehill from becoming a mountain if the manager is warned early enough to take corrective action. What is it worth to a manager to be warned of a business problem sometime sooner? What it is worth to the manager and to the company is the avoidance of a crisis. Often, it is too late to react to problems that are already out of control. This important advantage — gained from using problem finding — is very significant and real for the typical company today. It should be noted that in some situations, goal programming can be used to test financial results for achieving objectives for upcoming problems as well as future opportunities.

An integral part of problem finding is actively encouraging people throughout an organization to speak up. Top-level decision makers can bring to the surface many problems and important opportunities that might otherwise go unnoticed.

When decision makers need to mobilize people outside their direct lines of control to confront difficult problems and promising opportunities — as is almost always necessary — they need to build coalitions. Coalition building is particularly important for getting anything done. Sometimes, top-level decision makers have to make organizational changes, such as changing to a newer organization model, to guard against a potential disaster. Such changes always create winners and losers and generate overt and covert resistance. To prevail, they must be able to consolidate their supporters, neutralize their opponents, and persuade fence sitters to back the changes. That requires, in turn, that they be good at figuring out who wields influence inside and outside the organization, and then use that information and knowledge to build support and momentum for their cause. The bottom line is that if decision makers are unable to stay ahead of a potential disaster or to recognize important opportunities for an organization to pursue as they unfold, the organization is stuck in a reactive mode. As such, the organization is not a master of its own destiny (Watkins & Bazerman, 2003).

Connect Corporate Strategy to "Points of Wisdom" for Better Optimization of a Company's Resources

To assist in problem finding so that decision makers can solve future problems or pursue appropriate future opportunities, it is helpful for them to connect "points of wisdom" in their investigation. Addressing future problems and opportunities can center on three basic long-range corporate strategies. First, a company can invest available company resources in order for the business to grow and thereby strengthen its profitable position. Second, it can protect business strengths in order to maintain a strong position in a moderately attractive and mature market. Third, it can divest the business in order to exit from a weak position in a relatively unattractive market. Needless to say, connecting points of wisdom for better optimizing a company's resources differs in these three situations. In all situations, the discovery of new information, knowledge, and intelligence along with the optimization of a company's resources can be useful to decision makers in connecting points of wisdom to grow, protect, or divest a company's corporate strategy — from the short run to the long run.

To better visualize connecting points of wisdom for a corporate grow strategy — shown in Figure 3 — typically top management decision makers need to: (1) develop new products and services, (2) improve market share by expanding the sales force, and (3) increase manufacturing capacity. To connect these points in an optimum manner, a starting point is the utilization of *goal programming* to place the company's corporate strategy for a grow situation in its proper context. That is, a hierarchy of goals for these three areas can be modeled so that the organization's resources can be maximized to their fullest for what needs to be

Figure 3. The tie-in of goal programming with a venture analysis modeling approach for a corporate grow strategy

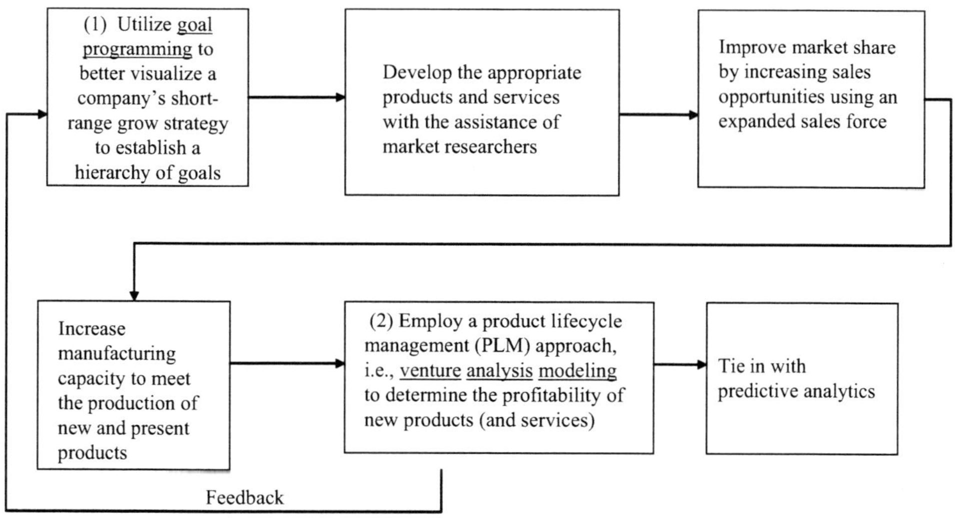

done. In turn, product lifecycle management can be employed to develop the profitability of new products (and services). That is, all sales and cost factors are evaluated to determine the profitability of the product (or service) over its lifecycle. As noted previously, this is known as *venture analysis modeling*. In effect, top management decision makers are making sound and wise decisions concerning the future profitability of a company's products and services. In a similar manner, the other two strategies, that is, protect and divest, can be optimized by utilizing appropriate mathematical and statistical techniques for connecting points of wisdom. The bottom line is that an organization's decision makers can make important connections so that their final decisions are optimum. Additionally, there is a need for measurement and analysis of results that are related to ongoing corporate performance.

To better understand connecting points of wisdom in Figure 3, that is, from goals for a company's products and services to the employment of a venture analysis modeling approach, it is helpful to develop a *cause-and-effect diagram*-as noted in the previous chapter. To establish a hierarchy of goals that will result in optimal decisions for the organization over time, top management decision makers must collaborate on the innerworkings of what products and services will be developed over the short to long run, what type of sales force will be needed, and the related manufacturing capacity for specific products. Essentially, goals need to be ordered that relate all of these three factors plus others deemed important. In effect, a root or basic element that starts the ongoing process needs to be determined as reflected in a cause-and-effect diagram in the short run. In turn, a hierarchy of goals in the short run can be formulated from the highest level to the lowest level that tie in with the medium-range and long-range goals.

Having developed the appropriate cause-and-effect diagram to assist in the proper formulation of this goal-programming problem, the next step is to *connect points of wisdom* such that optimal decisions can be forthcoming. In reference to Figure 3, the major points of wisdom that need to be connected are: (1) utilize *goal programming* to better visualize a company's short-range grow strategy to establish a hierarchy of goals and (2) employ product lifecycle management (PLM) approach, that is, *venture analysis modeling*, to determine the profitability of new products (and services). An effective connection between two basic elements allow a company's top management decision makers to be sent information, knowledge, intelligence, and optimization results about what is truly important to a company's overall operations over the short run so that wise decisions can be made. In turn, this vital input can be useful for the medium and long run for top management decision makers.

The last item shown in Figure 3 is a tie in with *predictive analytics* as discussed earlier in the chapter. Typically, a company wants to predict which customers will purchase the most product over their lifetimes. Predictive analytics is a form of knowledge discovery that enables the user to derive new insights or new information and knowledge from existing data and information. By contrast, past tools took information and churned it whereby information was presented in different ways. For example, the user can break out customer groups by gender and age and determine which groups are buying which products. Predictive analytics, on the other hand, lets the user perform analytics like predicting lifetime customer profitability — going beyond how much the customer has spent so far to how much the customer plans to spend over his or her lifetime with the seller. Predictive analytics is a leap beyond traditional methods because it offers faster processing speeds and more complex algorithms that also operate on top of existing knowledge discovery infrastructures. This means that the user can analyze more data and information over the short to long run. Additionally, predictive analytics are much easier to use than ever before (Harney, 2003).

For the first basic corporate planning function — *executive visioning* — in Figure 4, its activities are as follows. The first detailed activity — executive visioning relates external sources to internal factors thoroughly — focuses on top management getting the organization moving in the right direction for the times. Typically, this means relating to critical success factors analyses. The second detailed activity — relate turbulent times to executive visioning — allows top management to examine those current economic, social, and environmental factors that impact optimal decision making over time. The third detailed activity

Figure 4. Relationship of a company's corporate planning functions and detailed activities to connecting "points of wisdom" for their measurement and analysis

Basic Corporate Planning Functions (Figure 2)	Detailed Activities of Basic Corporate Planning Functions	Connect "Points of Wisdom" For Optimal Decisions in Corporate Planning	Measurement and Analysis Needed for Optimal Corporate Planning Decisions
Executive Visioning	(1) Execute visioning relates external sources to internal factors thoroughly	Recognize the need for starting with executive visioning	Projected income, costs, and profits plus ROI
	(2) Relate turbulent times to executive visioning	Examine grow, protect, and divest situations	Return on investment of capital projects
	(3) A broad-based approach to executive visioning may require changing the organization's business model	Examine the need for changing the organization's business model as times change	Profitability analyses
	(4) Provide feedback to executive visioning for possible changes	Utilize critical success factors for tieing in with visioning	Critical success factors
Corporate Mission and Purpose Plus Corporate Objectives and Goals	(1) Corporate mission provides the operational substance for a company	Who are the company's customers and what are their needs	Market research analyses
	(2) Corporate mission is related to a company's purpose for its existence	What useful need does the company provide for its customers	Market and investment analyses
	(3) Corporate objectives serve as guidelines for action over time	Objectives serve as a standard for measuring performance	Various performance analyses
	(4) Company objectives take the form of attainable, quantifiable, and measurable goals in the short run	Company goals are tied in with employee performance	Various financial analyses

Figure 4. continued

Basic Corporate Planning Functions (Figure 6.2)	Detailed Activities of Basic Corporate Planning Functions	Connect "Points of Wisdom" For Optimal Decisions in Corporate Planning	Measurement and Analysis Needed for Optimal Corporate Planning Decisions
Corporate Strategies and Programs	(1) Corporate strategies and programs are useful to allocate a company's resources in an optimum manner	Corporate strategies and programs can be set forth as specific objectives and goals	Critical success factors analyses
	(2) Corporate strategies emerge over time as a company responds to customer needs	This approach is useful in determining where funds are needed	Short- to long-term investment analyses
	(3) Effective strategies are developed through experimentation and design	The best marketing strategists see patterns taking shape	Marketing analyses
	(4) Examination of a company's strategies provides a basis for determining its corporate programs	Allocation of a company's resources provides a basis for financial plans	Return on investment of capital projects
Corporate Short- to Long-Range Plans	(1) Corporate long-range plans (five years and beyond) include projected products, manufacturing facilities, and personnel needs	Relates to connecting financial long-range plans in the future	Future market share and sales growth
	(2) Corporate medium-range plans (two to four years) include the above plus more detailed financial statements	Relates to connecting financial medium-range plans in the future	Medium-term market share and sales growth
	(3) Corporate short-range plans (current year) include monthly balance sheet, income statement, cash flows, and performance statements	Relates to revenue, cost, and profit interrelationships	Goal programming analyses
	(4) Feedback of current corporate short-range plans centers on profitability or lack thereof	Relates to identifying strengths and weaknesses	Venture analysis modeling

— a broad-based approach to executive visioning may require changing the organization's business model — has the capability of changing what products and services will be offered by the company. The fourth and final detailed activity — provide feedback to executive visioning for possible changes — centers on linking the results from financial measurement, such as return on investment of capital projects. This type of analysis can be an integral part of the balance scorecard approach. In a similar manner, the other three basic functions — corporate mission and purpose plus corporate objectives and goals, corporate strategies and programs, and corporate short- to long-range plans — which are explored in Figure 4 — can be developed in detail. In turn, these detailed activities can be connected to points of wisdom that allow top-level decision makers to judge soundly for making optimal decisions. The third and fourth detailed activities for corporate short- to long-range plans from Figure 4 relate to: (1) goal programming and (2) a venture analysis modeling approach set forth in Figure 3 for optimizing a company's operations for a corporate growth strategy.

Utilize a Venture Analysis Modeling Approach to Maximize a Company's Profits

An integral part of a venture modeling approach (as noted earlier) for an optimal KM/WM system centers on the development of appropriate products and services for a typical organization. However, the age-old question can be asked: "What products and/or services should be developed and financed?" arises. Hence, initially there is a need to do a thorough job of research regarding potential products and services for implementation. That is, the income from each product or service must be ascertained with a high level of accuracy. By the same token, there is a need to determine accurate costs for each product or service. The real focus is determining the profitability of each product or service over its lifecycle. One way of putting together the income streams, cost factors, and anticipated profits for each product or service is to set forth these values in the form of a capital budget analysis. Because a company's capital funds are limited, the individual capital projects are examined thoroughly so that each product or service can be evaluated in terms of its return on investment (ROI) and cash flows over their life. Generally, those capital projects that demonstrate high potential returns are given the green light for implementation.

In order to determine which capital projects are worthwhile to undertake, a venture analysis modeling approach means starting with a good understanding of what potential products or services that customers are looking for, and what new values can be incorporated into the products or services that best meet the customers' needs. In a similar manner, there should be consideration for all

Figure 5. The essential elements of developing a new product (or service) within an optimal KM/WM system environment

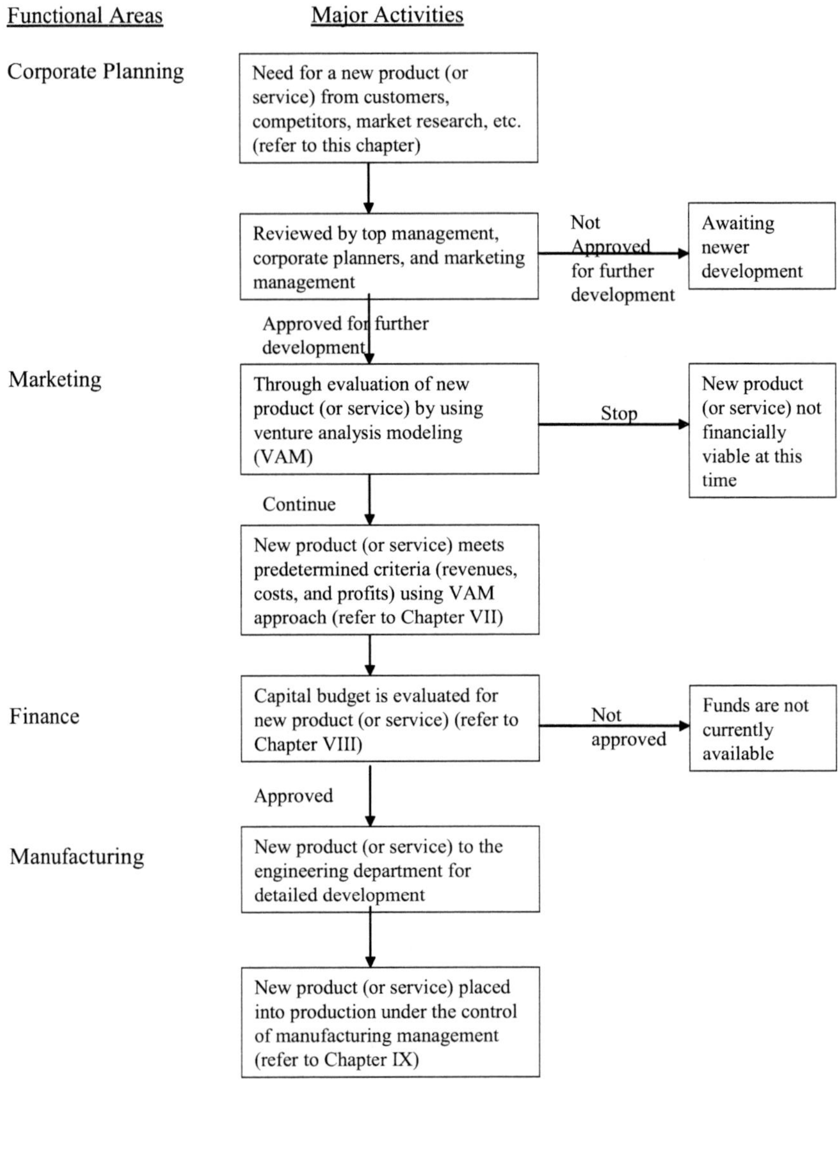

interested parties in the new products and services, that is, trading partners, stockholders, employees, suppliers, and the general public. From this view, this will result in a win-win situation for all interested parties. To better visualize the essentials of developing capital projects for the times, reference can be made to Figure 5.

As a starting point, the need for the product or service is identified through a number of sources — customers, company management, sales personnel, competing products or services, and market research. The need is reviewed internally by top management, corporate planners, and marketing management. In turn, the green light is given to market research to assure that the product or service is the one for the times. Next, there is a thorough evaluation of the income, cost, profitability aspects using a venture analysis modeling approach. The marketing aspects of VAM will be covered in Chapter VII while the essentials relating to finance and manufacturing will be covered in Chapters VIII and IX, respectively. If it turns out that the product or service is not viable for the marketplace, it needs to be terminated.

Continuing on per Figure 5, if the product or service is approved, adequate funds must be acquired. Due to the nature of the product or service, it may be necessary to undertake a capital project whereby external as well as internal funding is necessary. The subject matter of funding these capital projects is covered in Chapter VIII. Once the funds are obtained, it is generally necessary to engineer the product (or service) to meet the needs of customers. In turn, the engineering specification is turned over to manufacturing for acquiring parts and supplies from a company's trading partners. The essentials of manufacturing will be covered in Chapter IX. It is sufficient to say that the quality of the product (or service) is paramount to meet customer expectations. Fundamentally, a venture analysis modeling approach that makes use of current PLM software replaces the traditional approach to product development and manufacture found in the past.

To place this approach into perspective, the market research firm AMR Research Inc. (Boston) has set forth its VAM vision which includes the following components: direct materials sourcing for reviewing, selecting, and purchasing standard and custom parts; collaborative design for soliciting input across functions on what to build and how to build it; customer-needs management for gathering and analyzing customer and market requirements; and product-portfolio management for determining what resources to allocate to which projects based on a risk/return analysis. It is recommended that a company should make a VAM commitment that starts small and builds over the long term to help it better manage its present and future products and services for maximum profitability (Stackpole, 2003).

A typical example of the VAM approach can be found at EMC — a maker of data storage systems. EMC began VAM sometime ago as one of the legs in a corporate-wide strategic information technology rollout. The company tackled this approach in stages, but the stages kept the long-term game plan in mind. First, the company created a foundation of design data, information, and knowledge that could be shared among partners and contract manufacturers. Next, it linked design sources to key systems, including its Oracle Manufacturing application and other engineering tools. Looking forward, the company plans to open up those computerized data stores to partners and branch into other areas of VAM. EMC has reported that engineering change orders that used to take up to 22 days are now accomplished in hours, which means some engineers are 40% to 50% more productive than they were previously in certain phases of the design process. In total, EMC has benefited greatly from a VAM approach. It expects to benefit even more in the future by assisting company decision makers in short- to long-corporate planning of its capital projects over their life (Stackpole, 2002).

Evaluate Current Corporate Performance of Products and Services Using a BSC Approach

As noted previously in the text, especially in Figure 3, goal programming was utilized as a means of establishing a hierarchy of goals for a company's short-range growth strategy. In turn, a venture analysis modeling approach using current PLM software was employed to determine the profitability of new products and services over their lifecycles. Since a VAM approach represents a before the fact analysis, there is need to rate this performance with current corporate performance of a company's products and services. As discussed previously in the chapter, *balanced scorecard* (BSC) *software* keeps track of a company's financial progress in terms of a company's financial progress in terms of hard and soft measurements. These range from return on investment to customer satisfaction that needs to be managed to reach a company's desired objectives and goals, that is, profitable growth. A scorecard, for example, might graph customer service to determine if it is improving or deteriorating and, at the same time, tally product defects to determine if they are rising or falling and where. Essentially, a balanced scorecard approach is a set of regularly tracked measures that help an organization focus on and communicate strategic objectives and goals.

Currently, Robert Kaplan and David Norton recommend 20 to 25 measures as, the indicators of future performance, segmented into four categories (Kaplan & Norton, 2001):

1. *Five financial measures*, such as asset utilization rates, revenue per employee, and return on capital;
2. *Five customer measures*, such as the satisfaction and retention of customers;
3. *Eight to 10 internal measures*, such as the percent of sales from new products, new product introduction versus competition, and time to market of next generation products and services; and

4. *Five learning and growth measures*, such as employee satisfaction, percent of key-staff turnover, and percent of processes with real-time quality.

Each category contains measures critical to effective management as well as executive productivity and each is linked to one another and to an organization's strategy. Because the scorecarding process is top down, it starts with senior decision makers articulating specific strategic objectives and goals that are linked to corporate strategies and programs, including corporate short- to long-range plans (as discussed previously in the chapter).

As noted in the above four categories, financial objectives and goals are set first in order to demonstrate whether the strategy is being implemented effectively. Customer measures — the second category — reflect how customers view the company. Top management decision makers must explicitly state which customer segments it will compete in. Once financial and customer measures have been set, these decision makers must define what internal measures — the third category — for the company to excel at to meet or exceed customer expectations. This perspective is unique to the scorecard, as it forces the company to look internally at all new and existing processes that impact the customer's experience.

The final category — learning and growth measures — helps guide the organization through its investment decisions. A company's long-term value can be tied directly to its ability to innovate and learn. Specific objectives and goals that are measured should reflect this ability, for example, measuring time-to-market for new products. The scorecard also encourages organizational learning because it provides a feedback mechanism. There is a clear cause-and-effect relationship that can be diagrammed between the overall strategy and the measures. No measure is included that does not have a direct relationship to the stated corporate objectives and goals. By understanding these relationships, decision makers can adjust activities to improve the scores since a BSC approach provides a single view that aligns disparate divisions and activities of an organization. It looks at all activities and how they relate to each other and ultimately to the organization's objectives and goals. To be truly effective, the measures for the BSC approach must be consistent and mutually reinforcing whether they be hard or soft.

From an overall perspective, a balanced scorecard approach is usually distributed throughout a company's computer network and lets company employees across the entire organization be certain that they are talking about like items when they get together. If customer satisfaction is declining, sales, manufacturing, and research and development will all be reading the same score and thus decision makers will be able to tackle the problem from a common ground perspective. From an even larger perspective, a global company, for example, can create a common language that enables it to communicate its corporate strategy throughout it worldwide offices. It is an excellent tool for obtaining performance data and delivery feedback. A balanced scorecard approach allows this global company to achieve and demonstrate performance in a very short period of time. The bottom line using a BSC approach is a holistic view of a company's performance that assists top-level decision makers in judging soundly regarding about what needs to be done organizationally. Where past decisions have not been optimal, a balanced scorecard approach provides the essential hard and soft facts and figures needed to get the organization back on track in the future. Incorporating a BSC approach allows a company's top-level decision makers to gain a better understanding of how well the company is providing its customers with the products and services they need for the times.

An Optimal KM/WM System Application in Corporate Planning

The number of effective corporate planning applications within an optimal KM/WM system are few and far between at the present time. An important reason is that corporate-level decision makers have been and continue to be slow in adopting optimization techniques that can be employed to better facilitate the corporate planning process. The more popular optimization techniques today include goal programming, linear programming, integer and nonlinear programming, transportation, simulation, and assignment along with heuristic programming. Some of these techniques underlie the software found in Chapter IV as well as this chapter and the remaining chapters of the text. From the standpoint of tying together many operational areas of a company using these techniques, there is need to connect shared internal business processes that extend outside a company's boundaries. The bottom line is the improvement of a company's financial status over the short to long term by connecting business processes that reflect an enlarged view of wisdom for a company's decision makers. The maintenance of the organization is not left to chance, but is reviewed periodically to make sure that current business processes are attuned properly to changing or changed times. The downfall of the Enron Corporation and other companies attests to the failure of its corporate planning business processes in the form of the organization business model to reflect reality over the short term to the long term. In addition, these companies suffered from gross miss-management and greed.

One example of a corporate planning application is Salesforce.com, which is headed by Marc Benioff. It sells sales-force automation software that can be rented over the Internet for $70.00 a month. With 6,500 customers, Benioff has doubled the company's revenue over two years. Its main competitor, Siebel, sells a similar product that costs thousands of dollars per person upfront for software, servers, infrastructure, and management. Its sales have dropped. Currently, Salesforce.com is one of a host of companies that sell various sorts of enterprise software on a "by-the-drink" basis. Recently, the company's CEO Benioff has taken the concept further. The company has unveiled Sforce, a system that allows any software company or corporate customer to build its own software as a service product. Developers will use Salesforce.com's back-end services to build any business application they can think of. In addition, Salesforce.com offers dashboards that are targeted at marketing executives. Overall, these new directions are a recognition that there is need to connect points of wisdom at the corporate planning level, that is, what the customer really wants currently at a reasonable price, so that there is a win-win situation for all parties (Schwartz, 2004).

Summary

A beginning point for developing optimal KM/WM systems centers on corporate planning where there maybe need to reinvent the organization for optimal decision making. Conventional wisdom and an enlarged view of wisdom underlying corporate planning activities were addressed. In turn, the essentials needed in the development of an effective corporate planning model for optimal KM/WM systems were explored in some depth. In addition, a corporate planning model and its sub-models found in such an environment were set forth. Short- to long-range corporate planning was presented from the standpoint of executive visioning, problem finding, connecting points of wisdom, a venture analysis modeling approach, and evaluating corporate performance. Lastly, a real-world example of optimal KM/WM systems was presented. Essentially, this chapter on corporate planning serves as a basis for developing an optimal KM/WM system approach to growing an organization over time in the areas of marketing, finance, and manufacturing found in Chapters VII, VIII, and IX, respectively.

References

Begley, T. M., & Boyd, D. P. (2003, Winter). The need for a corporate global mind-set. *MIT Sloan Management Review*, 25-32.

Butts, M. (2002, June). At 92, Peter Drucker is a must read. *Business2.0*, 134.

Christensen, C. M. (1997). *The innovator's dilemma: When new technologies cause great firms to fail.* Boston: Harvard Business School Press.

Harney, J. (2003, January). Forecasting future trends from existing data. *Knowledge Management World*, 14-15.

Kaplan, R., & Norton, D. (2001). *The strategy-focused organization: How balanced scorecard companies thrive in the new business environment.* Boston: Harvard Business Review Press.

Norton, D. P., & Russell, R. H. (2003, March). The balanced scorecard: Aligning customer strategy with organizational resources. *ltol Magazine*, 39-40.

Schwartz, E. (2004, April 19). SalesForce.Com launches spring 2004. *Infoworld*, 22.

Stackpole, B. (2002, August). What's the next step? Product lifecycle management. *Managing Automation*, 18-21.

Stackpole, B. (2003, March). Demystifying PLM. *Managing Automation*, 16-22.

Watkins, M. D., & Bazerman, M. H. (2003, March). Predictable surprises: The disasters you should have seen coming. *Harvard Business Review*, 72-80.

Whiting, R. (2002, May 13). Business applications: Finance in focus. *Information Week*, 57.

Chapter VII

Optimal KM/WM Systems in Marketing

Issues

- To contrast conventional wisdom with an enlarged view of wisdom underlying marketing operations
- To examine areas that lend themselves to developing an effective marketing model
- To explore areas underlying marketing strategy and sales plans within an optimal KM/WM system-operating mode
- To set forth an optimal KM/WM system application that has been found to be successful in marketing.

Introduction

For sometime, companies have been living in a new economy where the customer is the focus. There has been a steady and inexorable slide of power from the producers of goods and services to their customers, both consumers and business buyers. Customers have become more demanding as increased competition and product commodization gave them more choices; smarter as they learned to exploit information resources, their own and Web-based; and both bigger and fewer following a wave of mergers. As a result, customers of today are immeasurably more knowledgeable and demanding than in the recent past.

Given this reality in the 21st Century, initially the chapter explores the new marketing power of today's customers, especially as found on the Internet. Next, an enlarged view of connecting "points of wisdom" in marketing is set forth in a number of management principles not found in the past. The essentials necessary to develop an effective marketing model for optimal KM/WM systems are presented along with an appropriate marketing model and its sub-models. Although many marketing areas could have been explored, the focus is on marketing strategy and sales plans that connect marketing strategy to "points of wisdom" for judging soundly as well as making this strategy an integral part of venture analysis modeling. In the final part of the chapter, an optimal KM/WM system application in marketing is given.

Customers Have New Marketing Powers Not Found in the Past

Since customers are now able to drive down prices, demand greater service, and expect constant innovation, they have new marketing powers not found in the past. Yet most companies have not figured out how to respond to this fundamental shift in the customer-supplier relationship. Many companies have published statements like "Customer satisfaction is our number one priority" in their mission statements. Yet these are mostly window dressing. A better approach is to employ optimal KM/WM systems for judging soundly about a company's marketing processes along with its related activities from the customer's point of view. Customers' needs for speed, accuracy, and flexibility cannot be met by overlays of conventional or poorly designed systems. The basic structure of the organization's business model, then, needs to reflect the customer's priorities, which is an integral part of optimal KM/WM systems in marketing today.

Today, there is need to present a single face to the customer. Many companies are organized into product-specific business units; a customer who orders products from several units typically has to contend with multiple sales personnel, uncoordinated deliveries, and a stack of invoices with inconsistent terms. Instead, the customer should be visited by one sales representative or sales team and should have one service person to contact with all problems; internal processes, systems, and databases should be standardized so multiple order-fulfillment activities can be pulled together to look like one.

Additionally, the company needs to implement and use measurements that reflect what is important to the customer. For instance, most companies define order-cycle time as how long it takes from when the order is placed until the goods are shipped. This is only a partial measurement. The customer cares about the time that elapses from the point that the individual recognizes the need to order until the time the goods are received. Similarly, the percentage of times a company delivers by the date it promised is much less important than the percentage of times it delivers by the date the customer actually requested.

When reference is made to the Internet and the World Wide Web, the Web site should always place the customer first. Walmart.com, for example, was revamped to meet its customer needs. Its focus in now on making all parts of its business constantly and relentlessly more nimble, more customer-centric, and more sensitive to the shifts and turns in customer taste and needs. The bottom line is that Wal-Mart is using the Internet to make all parts of its business — online, offline, or otherwise — focused not only on what the company does, but also more on what its customers want. Overall, marketing decision makers will find that connecting "points of wisdom" for what needs to be done organizationally over time will be made easier since there is only one focus, that is, the customer.

Focus on Customer Long-Term Value: "Return on Customer"

Because optimal KM/WM systems in marketing are helpful to grow the optimized organization over time, they need to stay relevant for changing times. Underlying relevance for the times is the recognition that *customer long-term value creation* is much more important than current earnings from a company's customers. A broad marketing perspective is the realization that "return on customer" (ROC) allows marketing decision makers to judge more soundly their decisions over time. More to the point, ROC allows a company's marketing personnel to calculate varying rates of value creation for different groups of customers, or even for individual customers. By understanding which groups or categories of customers can produce the most value will allow marketing decision makers to be more effective in prioritizing marketing and sales efforts along with research and development, production, distribution, training, and other key organizational activities. Customer centric companies include Dell and Best Buy. For example, Best Buy's approach centers on training its store-level employees to recognize and think about the different needs of five types of highly valuable customers, thereby allowing the customers to proactively satisfy their desires. The five groups that represent its core customers are: (1) affluent professionals; (2) focused, active, and younger males; (3) family men; (4) busy suburban mothers; and (5) small business customers. This focus on customer long-term value creation has proven to be profitable for Best Buy not only today, but also bodes well for its future (Peppers & Rogers, 2004).

Conventional Wisdom Regarding Management Principles Underlying Marketing

Over the years, a number of management principles based on information, knowledge, and intelligence of marketing operations have been developed. Essentially, these principles embody conventional wisdom when talking about effective marketing activities. An important principle revolves around the development and maintenance of core competencies of a company's products and services. Core competencies are the backbone of business development since they constitute the focus for strategy at the corporate level. They are the core of a company's operations in terms of its products and services. Generally, if a company is conceived of as a hierarchy of core competencies, core products, and market-focused business units only, then it will be able to survive in the long run. Cultivating core competence does not mean outspending rivals on research and development or becoming more vertically integrated. Rather, the focus is on doing what is necessary to ensure the continuing success of the company's basic businesses today and tomorrow. Essentially, core competence should grow and not deteriorate over time.

To assist marketing management in getting a handle on the changes affecting their products and services, there is a need for market research information gathered internally as well as information gained from market research forms. It should be noted that market researchers can also obtain appropriate information from magazines, newsletters, conferences, financial reports, trade and government organizations, and the Internet. No matter what the sources of market research used, it should be employed effectively. That is, it should be more than a mere guide for marketing managers to develop and support market-based decisions. Market research information over time can be assessed in terms of what has, and is, happening and what probably will happen to customers' needs for goods and services tomorrow. This broad-based approach to market research will give corporate management some insight into knowledge about the company's core products and services and where the company should be heading.

Typically, successful products today are the result of small and incremental innovations in design and manufacturing as well as marketing over the years.

Management must maintain a commitment to the product or service over a long period of incremental improvements. For example, Procter & Gamble persisted with research for 10 years to make Pampers affordable for the mass market. Because marketing management requires vision and persistence over many years against great odds, companies need to commit finances for the long run. Specifically, they need to address access to financial resources and willingness to use those resources. Enduring market leaders are companies that commit the required resources at the proper time.

Although newer information technologies allow companies to get a better understanding of their customers, there still seems to be a feeling among marketing managers that there is a straightforward linear relationships between *customer satisfaction* and *loyalty*. In reality, in markets where customers have a choice, the difference between the loyalty of completely satisfied customers and somewhat satisfied customers is quite different. In essence, competition can change somewhat quickly. Marketing managers who do not put forth their best efforts to satisfy their customers totally are at risk. In contrast, providing customers with outstanding products and services works in a positive way to achieve sustained customer satisfaction. For the most part, marketing managers who push excellence in satisfying customers are good at listening to their customers and in interpreting what customers with different levels of satisfaction are telling them.

In addition to the above, a number of other principles that are based on knowledge of customers have been developed over time. A most notable source is Profit Impact of Marketing Strategies (PIMS), a computerized approach for planning marketing strategy run by the Strategic Planning Institute. It is a data pool of information on the marketing experiences of its members (Buzzell & Gale, 1988) Several hundred corporations submit data annually on about 3,000 of their business units, each of which is a distinct product-market unit. Each member provides PIMS with the most intimate details on matters such as its market share, investment intensity, product quality, and total marketing and R&D expenditures. Through computer simulation, the company can test its own market strategies against the real experiences of hundreds of comparable companies, including competitors. What it receives are answers to questions such as these: What is the normal profit rate for a business or a product line given its combination of circumstances, and why? If the business continues on its current track, what will its future operating results be? What will happen to short- and long term performance if certain strategic moves are made? What changes will create the best profits or cash returns?

One of the findings that has emerged from the PIMS computer models of real-life experiences of its corporate members is that there is a set of operating rules that governs all businesses. Some 37 factors — including market share, capital intensity, and vertical integration —jointly explain 80% of the success or failure of any business; only 20% of a business's return-on-investment can be attributed to factors that are unique or special, such as the quality of working relations. Other PIMS findings indicate that a larger market share is strongly linked to higher profits and that product quality is also very positively related to return on investment. When product quality is low, it does not pay most companies to have high marketing expenditures. High research and development spending hurts profitability when market position is weak, but increases it when market share is high. In these situations, copying competitors' products rather than inventing new ones is probably the best bet. In addition, higher investment intensity, that is, ratio of total investment to sales, is associated with lower ROI, and high investment intensity combined with low market share is an ROI disaster.

All in all, what a member wants from PIMS is to find out what it will cost to make a particular strategic move and how much better off the business will be afterward. For example, consider return on investment, which PIMS considers one of the best measures of how a business is doing. The PIMS models can forecast how much ROI for a business line will change because of a strategic move involving more marketing, research and development, capital equipment buildup, or whatever — both what the return on investment will be immediately following the move and what it will be several years in the future. Principles that have emerged from the PIMS-computer models of the real-life experiences of its corporate members are highly regarded by marketing decision makers.

An Enlarged View of Connecting "Points of Wisdom" in Marketing

The management principles set forth above that are an essential part of a typical company's marketing operations are not the end point. Optimal KM/WM systems require a refocusing on the customer — as noted earlier in the chapter. For example, Henry Ford (with his inexpensive mass produced Model T car), George Eastman (with his low-priced camera to take one's own pictures), and Sam Walton (with his new approach to mass merchandising of consumer products that are purchased everyday) all focused on the customer from a new and fresh viewpoint. In reference to Sam Walton, the founder of Wal-Mart, he recognized the force behind the supply chain as it is known today. He brought about this shift in market savvy through his accent on customers. Walton had taken a hard look at the customer and saw quite clearly that the actions of the customers in his stores were the keys to expanded business. He helped create a system whereby as a product left the shelf through customer choice, the store instantly monitored the sale and was able to initiate a replacement of the product automatically. In addition, he connected in a most important way how the customer and the supplier interact, whereby the store became the enabler of this process. Sam Walton's ability to judge soundly from the beginning about this basic process has enabled his corporation to become the largest one in the world (Malone, 2003).

In light of this discussion, a most important newer management principle in marketing centers on tying together a company with its suppliers and customers by integrating processes and information flows with theirs. The real advantage is not in cost savings, but in interlinking businesses so they can help one another better. Wal-Mart, for example, has not battled with its suppliers, but rather it has joined forces with them to bring products to market faster or to achieve other mutual advantages. Wal-Mart, which started out with electronic data interchange in the early days, has moved far beyond that technology to a system developed in-house through which the company openly shares information with its suppliers. In fact, that system, called RetailLink, goes so far as to put each supplier roughly on an equal footing, in terms of information, with Wal-Mart's own internal buyers and financial analysts. In effect, Wal-Mart has moved its integration up and down the supply and demand chain to better meet customer needs. This principle plus many others, that center on what needs to be done by marketing decision makers to grow the optimized organization over time, are found in Figure 1.

Going beyond the integration of customers and suppliers, there is need for companies to go further. Currently, some companies have figured out a way to outdistance their rivals through custom-focused strategies that are virtually imitation-proof. More specifically, this means capturing value in customer interactions, which goes beyond just better products and services. This can take the form of: (1) unlock the economics of interaction by lowering customer costs and reducing the risks faced by customers, (2) simplify the route to benefit from new business models, (3) integrate activities with customers to lower operating and transaction costs, (4) bring related offers and complementary systems to current customers, and (5) focus on forming the future for customers in order to reduce the lifetime cost for customers. By being the link between sellers of complementary products and services and their customers, there can be an increase in the number of touch points between parties in order to increase loyalty among them. Thus, tremendous opportunities for creating values with a company's customers are extremely helpful in establishing long-term advantages. A broad-based view of wisdom helps companies outdistance their competitors over time (Vandenbosch & Dawar, 2002).

While conventional wisdom states that creating a brand is about differentiating the product or service, an enlarged viewpoint of wisdom starts with brand positioning by establishing a frame of reference. This signals to customers the

Figure 1. Newer management principles help marketing decision makers get an enlarged view for connecting "points of wisdom" about what needs to be done over time

- Successful marketing efforts integrate a company's operations from its suppliers through to its customers whereby feedback is an important part of the entire integration process. The capability to move integration up and down the supply chain is a must today to better meet customer needs.
- Tremendous opportunities for creating value with a company's customers are extremely helpful in establishing long-term advantages. A broad-based view of wisdom help companies outdistance their competitors over time.
- Important advantage comes from utilizing brand positioning whereby a frame of reference is established. Brand positioning requires that it be consistent at any point in time which is recognized by the customer.
- A venture analysis modeling (VAM) approach which has been an important topic throughout the text can be integrated with collaborative product/service commerce (CPC). A CPC foundation provides the single logical product data set that supports concurrent activity based on shared visibility of that data.
- The capability of today's information technology to sort through massive amounts of information, knowledge, and intelligence to target small groups of highly responsive customers allow large companies to own market niches the way small companies do.
- Senior marketing managers and their staffs need to spend several days in the lives of their key customers since there is no substitution for personal knowledge of the marketplace. Creativity can bring new insights and expertise to their customers' needs and problems.
- The underlying marketing framework today centers on creating, meeting, and improving customer demand. Customers own their product and service suppliers in every way today and will do so even more tomorrow. If customers do not show up and buy, supplying companies are dead and they know it.

goal they can expect to achieve by using a particular brand. In some cases, the frame of reference is other brands in the same category. Next, there is a need to think through the points of parity that must be met if customers perceive the product or service a legitimate and credible player within that frame of reference. Last, the points of difference should be feasible. Although customers may find low prices or free delivery attractive as points of difference, all too often these features compromise profitability and are easily imitated. Developing an effective brand position goes beyond delivering the frame of reference, points of parity, and points of difference. It also requires that these elements be consistent at any point in time. The bottom line is that an enlarged view of wisdom ensures that these elements do not contradict one another, but are complementary. As a brand ages, the challenge for marketing decision makers is to make sure it stays up to date and in touch with the customer's changing needs. Typically, frames of reference, points of parity, and points of difference are moving targets (Keller, Sternthal, & Tybout, 2002).

To assist in connecting points of wisdom within an optimal KM/WM system-operating mode, a venture analysis modeling (VAM) approach integrates multiple facets in the lifecycle of a product — customers, engineering design, sourcing, marketing, manufacturing, and suppliers — from the outset. Essentially, there is need for a product lifecycle management approach that allows risk to be managed and controlled with clear go/no-go decisions for a new product or service. Companies need a modular process such that workflow segments within each stage can be configured to fit the product or service in question. A collaborative product/service commerce (CPC) foundation provides the single logical product data set that supports concurrent activity based on shared visibility of that data as well as supporting the workflow that defines each contribution and when it is required in the process. Marketing professionals have a critical role to play in product lifecycle management because supplier management and supply-chain improvements often make up a large part of the business case for CPC. From this perspective, it is recommended that a CPC approach be tied-in with lifecycle management of a company's products and services (Williams & Stemper, 2002).

Although a product lifecycle typically follows a period of introduction, growth, maturity, and decline over time, marketing decision makers can position their products and services in new and different ways. They can shift products or services lodged in the maturity phase back to the growth phase. That is, products and services can be changed by how customers mentally categorize them. Effective product and service positioning strategies include reverse positioning (strip away traditional product or service attributes while adding new ones), breakaway positioning (associate the product or service with a completely different category), and stealth positioning (offer a new approach to the product or service by cloaking its true nature). In effect, companies can utilize these techniques to redefine itself in terms of traditional boundaries for products and services (Moon, 2005).

To assist large companies in meeting marketing challenges from smaller companies, newer information technology such as knowledge discovery is helping them sort massive amounts of information to target small groups of responsive customers almost automatically. For example, a large supermarket chain can use point-of-sale scanning and frequent-shopper programs to build a sophisticated customer database. Extensive customer information enables this supermarket chain to stock each store according to the preference of local customers at the same time that it centralizes its buying process. Also, such customer purchasing programs can be used to target individual consumers and, eventually, individual households. Overall, massive amounts of information can be culled to discern trends and patterns that are useful in targeting small groups of highly responsive customers.

To assist marketing management and its staff in getting a handle on important factors affecting a company's marketing operations, the principle of spending time with customers goes a long way toward understanding their point of view. This recognition about the customer's importance suggests that the customer is more than the next step in the distribution chain. Senior marketing managers and their staffs need to spend several days in the lives of their key customers. There is no substitute for senior executives' instincts, imagination, and personal knowledge of the marketplace. Creativity on their part can bring insights and expertise to their customers' needs and everyday problems. Hence, the strategic planning process for all of a company's marketing areas can be improved by making the customer the focus of investigation. In turn, knowledge that is learned about a company's customers can help provide an underlying framework for an optimal KM/WM system.

An underlying framework for the newer management principles in marketing center on demand, that is, how to create and improve it. Today, marketers do not control customers, but rather customers control marketers. Customers own their product and service suppliers in every way today and will do so even more tomorrow. Essentially, the customer has all the power in the marketplace and marketing organizations are seeing what little control they had slip away. This is particularly true for any manufacturer who sells to or through Wal-Mart, Home Depot, or Walgreens. They will quickly state that the customer controls that system from the other side of the table and one gets the same answer from the retailer. No matter how skilled Starbuck's becomes in locating new stores or how much faster Dell becomes in assembling computers to order, if customers do not show up and buy, these companies are dead and they know it (Schultz, 2003).

Development of an Effective Marketing Model for Optimal KM/WM Systems

The development of an effective marketing model for optimal KM/WM systems needs to take into consideration the preceding newer management principles found in marketing. Marketing decision makers must make customers their partners. Both customers and partners need to share information, knowledge, intelligence, and optimization results in a collaborative arrangement. In essence, collaboration brings benefits of such integration to all participants. Related to the development of an effective marketing model are the following items: (1) examine an enlarged view of market research and analysis, (2) focus on a venture analysis modeling (VAM) approach, (3) view customer relationship management (CRM) from a broad perspective, and (4) employ marketing software useful in optimal KM/WM systems. All of these topics are discussed below.

Examine an Enlarged View of Market Research and Analysis

To satisfy and, at times, exceed the expectations of a company's customers, it is necessary to take an enlarged view of market research and analysis. But how does a company know what customers want? Typically, because of the way most traditional market research and analysis is conducted, it has fallen short of its important objective. At the core of the problem is the practice of using market research to confirm that a decision already made is the right decision rather than using market research and analysis to identify alternatives and to support the process by which the best alternative is chosen. An enlarged view of market research and analysis also includes exploring new opportunities that require in-depth insight into the customers' needs, lifestyles, and aspirations.

Over the years, market researchers have found that although consumers give accurate responses when asked whether they would try a product initially, contrary to expectations, consumers were not very good at predicting whether they would buy new products or existing goods. In fact, the market researchers have found that people are not generally reliable predictors of their own long-term purchasing behavior for any type of product, new, or old, durable or not. Asking consumers if they intend to buy a computer in the next year of two, for instance, will not generate responses that match up very accurately with actual sales. Basically, measures of intentions are better predictors of how many people will buy the product in the short term than how much any of them will buy over the long term or how much will be eventually sold.

A most important consideration in market research is asking consumers about their purchase intentions in a context that is as close as possible to the one they will actually be in when buying the product in question. To that end, the market researchers recommend that marketers first ask a number of background questions aimed at making respondents sensitive to their own behavior. For example, before springing a new flavor of low-fat ice cream on a sample of consumers, marketers should ask them how often they buy ice cream, what other things they eat for dessert, and whether they have been trying to cut down on fat. Marketers should even make consumers aware of the other brands of ice cream they would be giving up by selecting their product. Typically, this is how consumers really make decisions in a supermarket, that is, they do not buy in a vacuum. They pick a product off the shelf that is sitting next to others.

Going beyond asking the right questions in the right way is realizing the limits of what the answers can reveal. If marketers really want to know how much of their product consumers will eventually buy or whether consumers will keep coming back for more after the initial purchase, they need to understand that it is not enough simply to ask. Overall, purchase-intention surveys can take market research and analysis so far. Marketers would also do well to observe consumers in real buying situations for better market results.

Focus on a Venture Analysis Modeling (VAM) Approach

A *venture analysis modeling* (VAM) approach is a way of analyzing opportunities for new products and services that encompasses such techniques as probability, decision theory, and the time value of money as well as mathematical modeling. Venture analysis looks at the various periods of a product's life, that is, its introduction, initial growth period, introduction of competitive products, market saturation, market decline, and removal from the marketplace and replacement by newer versions or new products. Because venture analysis is a massive system of gathering, relating, appraising, and projecting all data pertinent to a complete business venture over its lifecycle, this mathematical technique stores many kinds of information. All costs involved in the product project are developed. Manufacturing costs include raw materials, direct labor, depreciation, and overhead. These data are modeled for each step of the production process. For the pricing and promotional effort, product prices are estimated at various desired levels to determine the optimum price. All promotional costs involved in marketing the product are broken down by media sources selected and projected. Research-and-development costs plus general-and-administrative costs are developed to make the data complete.

An important part of the venture analysis model is knowledge about the proper determination of prices at various stages over the lifecycle of the product by marketing modelers. Utilizing a pricing approach within venture analysis, the model relates a myriad of price, advertising, personal selling, and sales promotion combinations for the product under study and pertinent facts about competing products. Essentially, the venture analysis model hypothesizes about the degree to which competitors will react to a price change and in what form this reaction will occur. It analyzes using an appropriate statistical and/or mathematical model what blend of marketing decisions will go best with a given price. In turn, it determines what effect the given price will have on the sales of other products in the product line. The pricing aspects of a new product, then, are included in the venture analysis model along with the prices charged by competition. However, it is up to the marketing managers to review these prices and make appropriate adjustments based on their knowledge for reasonableness as well as their experience and judgment. Otherwise, unrealistic prices will be used to evaluate a product or service over its lifecycle. In turn, the product or service will be priced unrealistically in terms of its competition.

View Customer Relationship Management (CRM) from a Broad Perspective

Due to the impact of the Internet and the World Wide Web, the basic tenants of where the customer is primary (i.e., the customer is king) and knowing the customer are dramatically shifting and becoming more sophisticated. When customers visit a company's Web site, they are expected a deeper and wider range of information and knowledge access and delivery. More often than not, the information and knowledge expected will be more sensitive and precise than maybe a company is comfortable providing. As companies have learned, every new bit of information and knowledge shared with customers will cause them to want more. Certainly, *customer relationship management* (CRM) is all about companies providing their customers with the personalized information, knowledge, plus resulting intelligence and optimization results that is needed to run their businesses. Companies win not only by accelerating, automating, and optimizing their own business systems and decision-making environments, but also by delivering facts and figures that helps customers and suppliers accelerate, automate, and optimize their decision-making processes. The bottom line is that each decision has impact on other decisions such that there is a ripple effect throughout multiple organizations.

In the past, there was a dependence on personal contacts. However, current total customer relationship management is much more than just marketing, sales, or customer service. It includes the customization of a Web site based on a customer's preferences, characteristics, or previous behavior as well as versioning a help desk in the customer's native tongue. In addition, it includes offering or denying house credit at point of service based on previously known payment history. Using analytical measures such as product affinity and propensity scores, channel and media preferences, and attrition risk scores at the point of sale or service, CRM helps define for a company its best customers. In totality, total customer relationship management covers the entire sales and service experience and where improvements should be made.

Customer relationship management essentially combines a set of business disciplines (finding out who a company's customers are and what they want) with technology (storing that information in a database and using specialized software to sort it out). In turn, that marketing intelligence can be used to find ways to make more money from customers as well as improve the quality and efficiency of serving them better. A customer relationship department serves as a liaison and advocate group for customers, that is, ensuring that all areas of the organization are prepared for customer interaction and optimization and that customer relationship strategies are successfully implemented across the organization.

Users today are seeing a wave of new versions and releases as CRM vendors expand and improve their products' Web-based features. Yet despite these improvements, various analysts and some vendors claim that Web-enabled solutions (which are a hybrid of client/server systems and the Web) are not enough. Truly, competitive CRM solutions must be Web-based, that is, these solutions utilize the Web as their platform. Traditional Web-enabled CRM systems have been oriented around managing activities inside a company, whereas e-commerce is oriented around customers. The real issue is building a long-term relationship with customers. And paramount to e-commerce is the realization that as far as customers are concerned, moving to a competitor is only a click away.

In a Web-based operating mode, all information, knowledge, intelligence, and optimization results about customers, products, marketing, positioning, collateral, and competitors reside on the Web. CRM focuses on the analysis of available business intelligence, demographic data, lifestyle information, and purchase behavior, thereby creating detailed electronic profiles of individual customers so that companies can market more effectively. In addition to Web services, the CRM industry is centered on real-time CRM. The broader acceptance of the Internet and the World Wide Web as a communication, content, and application medium will add to an even wider spread of CRM. Businesses of all types are increasingly relying on the Internet for managing their ongoing, real-time customer relationships. The wireless revolution, which includes the explosion in Wi-Fi hot spots and 3G wide area networks, is also extending the range of CRM for businesses. Hence, when it comes to real-time CRM, the impact on private and public organizations will be profound, and will significantly improve the way these organizations conduct business (Goldenberg, 2005).

Employ Marketing Software Useful in Optimal KM/WM Systems

A number of marketing software packages for optimal KM/WM systems are helpful to decision makers in order to discover their customers' needs and wants over the short to long term. Many times, these software packages include broad-based techniques, such as goal programming, linear programming, Markov analysis, game theory, and simulation, which are useful for shaping an overall marketing strategy and its related plans. Typical marketing software packages found in optimal KM/WM systems center on the following: (1) optimization, (2) behavior-mapping, (3) product lifecycle management (PLM), (4) knowledge discovery (data mining), and (5) CRM analytics. Each of these software packages is set forth below.

An important concept underlying the above software packages is the concept of customer "lifetime value" to design customer-based strategies and to measure their return on investment. This concept was covered earlier in the chapter and was referred to as "return on customer." The lifetime value of a customer is the present value of all future profits generated from that customer. In a simple theoretical example, if the customer-retention rate is 100%, that customer generates a certain profit margin for each period measured. In this case, the lifetime value of this customer is simply the present value of the future income stream. This is identical to the finance approach of valuing perpetuity. Although it is possible to build sophisticated models for estimating customer lifetime value, a useful simple estimate is the annual profit margin of a customer multiplied by a margin multiple. This multiple typically ranges from one to five, and in most cases, an optimistic estimate for this multiple is four. In other words, if a customer provides a margin of say, $100.00 per year, the lifetime value of this customer is about $400.00. This simple idea can lead to significant insights for decision making, such as how much to invest in customer-data collection and analytics (Gupta & Lehmann, 2002).

Under *optimization software* found in Chapter IV, typical vendors that integrate optimization techniques into their software packages include Aspen Technology Inc., CAPS Logistics, Inc., i2 Technologies, Ilog Inc., Manugistics Group, Siebel Systems Inc., and Triology. Still other software useful in tactical optimization of marketing centers on the interactive world of the Internet for more in-depth, complex customer profiles. Fortunately, newer software is emerging that can better meet e-commerce customer relationship management and marketing needs. For example, *behavior-mapping technologies* construct in-depth profiles of each customer based on their interactions with the business across all communication channels, including the Internet, ATMs, and call centers. These technologies provide a comprehensive view of customers, thereby enabling businesses to view their behavior continuously as they evolve. Instead of storing customer information in ways that make retrieval and analysis unwieldy, these technologies continuously distill and interpret behavior into current, multidimensional maps of each customer that enable businesses to respond intelligently and instantly to customers' needs. Equally important, this approach detects when a customer's relationship with the business is beginning to change for better or worse — enabling an immediate response to either save the relationship or expand their business.

Another measurement approach is venture analysis that looks at the profitability of a product or service over its lifecycle. This approach was mentioned earlier in the chapter. It is sufficient to say that both lifetime value and venture analysis center on maintaining an optimized organization over time. Currently, venture analysis employs *product lifecycle management* (PLM) software. Underlying PLM is the realization that some kind of integration strategy is required to ensure the much-needed connections to key enterprise and legacy systems. Most of the leading PLM vendors have struck partnerships with middleware vendors to deliver on that requirement and/or have an open architecture on which to build PLM products. Others are teaming up for specific integrations like the ones resulting from the Agile/PeopleSoft partnership or the one between PTC and Siebel Systems Inc. Most of the big technology suppliers are teaming up with enterprise application integration (EAI) vendors or providing their own to facilitate the integration effort. There is a natural integration with ERP, CRM, and SCM. PTC, for example, has a deal with TIBCO Software Inc. to provide packaged integration solutions between PDMLink and leading ERP packages, including those from SAP AG, Oracle Corporation, and J.D. Edwards & Company. IBM Corporation and Dassault Systems S.A., for their part, offer IBM's WebSphere, MQ Series, and Cross-Worlds flavors of middleware as well as integration services from IBM Global Services to address the need. Currently, PLM software has a number of success stories from companies in the aerospace and automotive sectors. These successes have centered on collaborative marketing and engineering efforts. Companies tend to focus downstream on how to create, design, and get products moving into manufacturing. From this broad perspective, PLM marketing efforts start the company's efforts to get new products or services into the flow of a company's daily operations (Stackpole, 2002).

Knowledge discovery or *data mining* allows the user to slice and dice a company's database, looking for patterns in order to provide the most useful "nuggets" to guide marketing efforts. As discussed previously in the text, data mining solutions help a company provide sophisticated analysis functionality for predictive modeling, trend analysis, application scoring, and customer segmentation. The software helps identify the most profitable customers, detect and define customer attrition trends, determine optimal timing for product rollouts, model customer buying behavior, and discover previously unrecognized patterns in customer data that lead to new marketing opportunities. Today, optimization is considered the most advanced stage of knowledge discovery usage and offers a large potential payoff. Essentially, it uses sophisticated data-mining algorithms that sift through data volumes to discover patterns that may be too subtle for humans to distinguish. The software applications can then automatically apply the insights to optimize customer interactions. In other words, by tailoring recommendations and promotions to the preferences of specific customer groups, the user can actually change customer behavior by up selling them to a higher-priced product; cross-selling to additional, related products; or even down selling to a lower-priced product in cases where the customer is abandoning a potential purchase because its price is too high. These recommendations, based on data patterns, can produce immediate payoffs in the form of increased sales (Fayyad, 2003).

Going one step further, *data-mining visualization* is a way to visually demonstrate the hidden patterns and relationships that often result from data-mining activities. Instead of looking at a Microsoft Excel spreadsheet or a list of numbers or scores, decision makers can literally see how data relate to other data, usually in the form of a bar graph, pie chart, or, as in the case of visualization vendor 3DV8, a swirling solar system. The biggest difference between the output of standard data mining and data visualization is the results. Instead of continuously analyzing, the data visualization tools can point out only the changes in the data or give decision makers a new perspective on the information and knowledge.

Various types of analytics can be used in marketing. To assist marketing decision makers, *CRM analytics* provide them with a number of key performance indicators (KPIs) which gives them a "snapshot" of who are their most important customers. The ability to provide a "unified view" of customer information is an important function of CRM analytics. If the application allows the marketing manager to see integrated information and knowledge that comes from legacy systems, the individual can view the customer's entire history, all the way back through the supply chain. When marketing makes this kind of investment in customers — and they are aware of it — these customers will be less likely to leave based on price alone. Additionally, the Internet has unleashed the potential of CRM applications, delivering more value for each company, and sharpening the tools to help enterprises understand, serve, and anticipate customer's demands. The advent of Web services should propel CRM even further toward achieving its ultimate goal of boosting sales by giving customers the services and terms that keep them coming back for more.

An example of predictive analysis software is SPSS Predictive Marketing that uses pre-built models to help marketers answer questions such as which customers are likely to respond to a sales offer. The software can be used to identify customers who are likely to abandon a business, are more likely to respond to a new offer, or have the potential to generate greater profits for a company. SPSS's Clementine software can be used to build additional predictive models.

Marketing Model and Its Sub-Models Found in Optimal KM/WM Systems

Before a marketing model and its sub-models found in an optimal KM/WM system can be developed successfully, there must be a recognition that an effective approach to marketing must start with creative thinking as related to problem finding and solving (refer to Chapter II). In the development process of a marketing model, it is necessary to employ a CPC approach as set forth above that centers on adding unique values for customers. Additionally, there is need to evaluate a company's products and services over their lifecycles. Current research indicates that about 46% of all resources spent on the development and launch of new products is wasted on products that either never make it to market or fail upon arrival. Of those products that do make it to implementation, between 60% and 90% do not meet customer expectations, depending on the industry. The design goals of lifecycle management for products and services, however, are to improve these odds by employing a through-life focus on products and services that includes design, production ramp-up (time-to-volume), introduction to the market, in-life yield management and planned upgrading, and end-of-life management (Williams & Stemper, 2002).

A logical starting point — as shown in Figure 2 — for an effective *marketing model* and its major sub-models is *executive visioning* which centers on relating external sources to internal factors. To tie in these sources and factors, initially this means that marketing management must develop *marketing strategy and sales plans sub-models* (Figure 2) such that market demands for a company's products and services are adding vales for its customers, suppliers, and employees. To accomplish desired marketing objectives and goals that reflect unique customer values, marketing management must develop short- to long-range marketing strategies and related programs. Today, these strategies and programs must take into account engineering costs out of products/services so that savings can be passed on to a company's customers. Similarly, there is need to engineer time and costs out a company's processes and activities that add to its bottom line. Overall, the focus of this first sub-model is to instill in marketing personnel the necessary wisdom to detect when there is a need for changing the marketing direction of the organization.

To assist in relating external sources with internal factors and realizing unique values for all of a company's partners, marketing management needs to employ a comprehensive *market research and pricing sub-models* (Figure 2) approach that recognizes the increasing speed of product/service evolution. At the same time, marketing management must keep in mind the necessity to create and distribute new products/services that really add unique values for customers. As

Figure 2. Relationship of the marketing model and its sub-models to the other functional models within an optimal KM/WM system

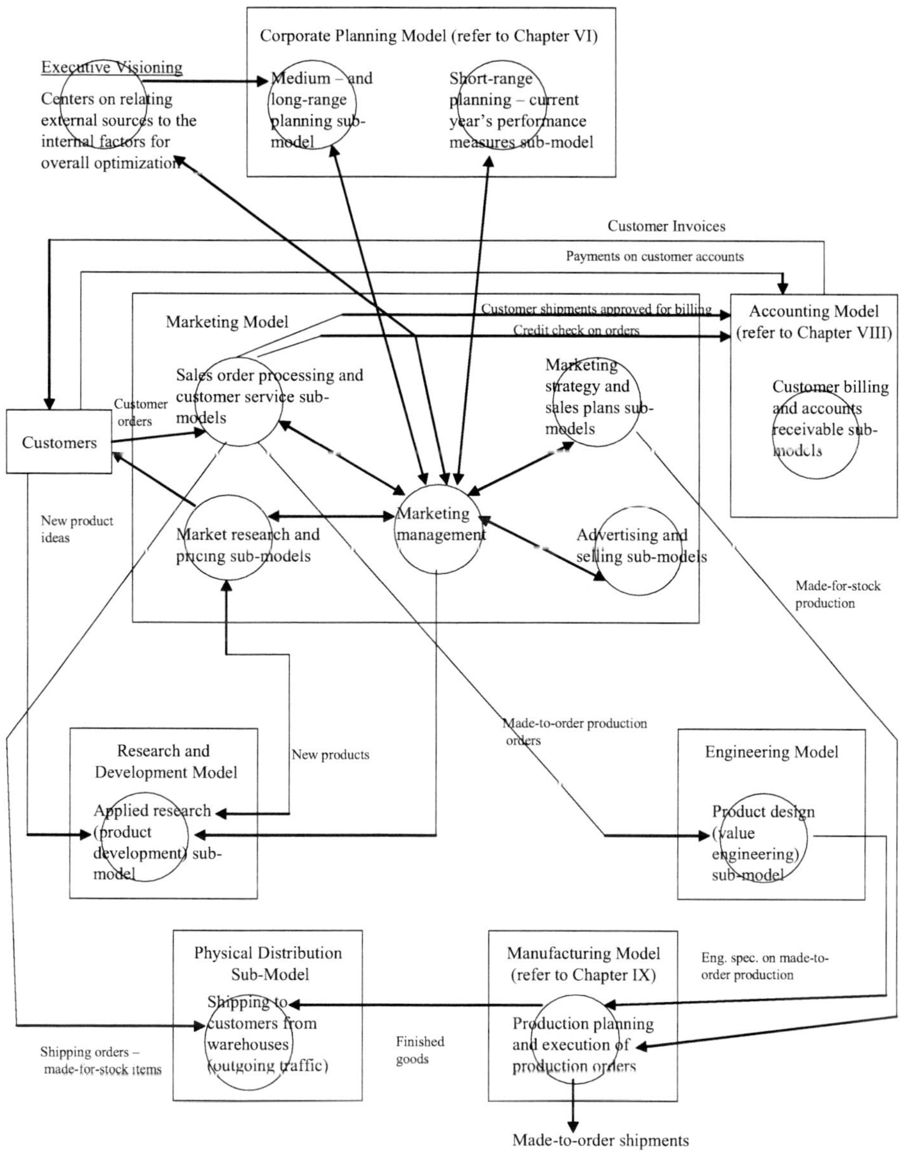

discussed in a previous section, the utilization of PLM and venture analysis that recognizes the need to vary prices over the life of a company's products/services can be helpful to maximize a company's profits over the long term. Hence, the wisdom of marketing management for appraising new products/services along with the maintenance of current products/services is reflected in a company's marketing strategy and sales plans.

There is a need to tie in the marketing strategy and sales plans sub-models with an effective *advertising and selling sub-models* approach (Figure 2). In terms of advertising, a media model that reflects a total advertising budget for each media and exposure rates is needed. The net result is an allocation of the advertising budget to various advertising media. Advertising of products/services leads to customer orders via the Internet, fax, mail, and so forth. Generally, a wide variety of activities is involved in a company's sales order processing. Marketing management should have the capability of evaluating online sales orders and activities over time so that a company's marketing decision makers can distill a certain amount of wisdom from the company's selling experiences.

A most important marketing activity centers on the *sales order processing and customer service sub-models* (Figure 2). Once the sales order is received, the first major activity, aside from detaching proceeds, is the credit check, that is, is the customer capable of paying the ordered amount? In those cases where the credit of the customer is in question according to the accounting department, the order is sent to marketing management for further checking and a decision. In most cases, credit clearance is obtained and the order proceeds to the next step to see if there is need for product design (applies to special orders). If engineering is required, the order departs from the normal sales order processing flow and ultimately the special order or made-to-order items are manufactured and delivered to the customer. On the other hand, made-for-stock items are forwarded to physical distribution for future delivery to customers. In conjunction with sales order processing, the level of customer service is considered. If the order can be filled from inventory or from current production, the order processing cycle continues with the development of shipping documentation and the scheduling of warehouse withdrawal and transportation. When the order cannot be filled from either source, the customer is notified through the marketing department. Hopefully, the customer will allow the order to be backordered and filled at a later date. If not, the customer will probably use an alternative source and the order will be canceled.

Related to the above marketing sub-models is the *physical distribution sub-model*. As shown in Figure 2, finished goods are the output from the manufacturing model. They can be shipped directly to the customer upon the receipt of customer orders or stored in company and non-company owned warehouses. Periodically, which includes hourly, there can be an online evaluation of finished goods and warehouse shipping schedules by marketing and physical distribution management. Similarly, there can an online evaluation of out-of-stock conditions and back orders by management to determine where the company is falling down in terms of meeting customer needs. A vast experience of a company's current distribution channels and forthcoming ones will assist a company's marketing decision makers in judging soundly about the physical distribution needs for its customers, its suppliers, and the company itself. Overall, the activities of the major marketing sub-models are interconnected to physical distribution. All of these activities provide a way of connecting "points of wisdom" for evaluating a company's critical success factors (CSFs), key performance indicators (KPIs), and financial ratios thoroughly.

Marketing Strategy and Sales Plans

A marketing strategy to accomplish short- to long-range sales plans for an optimal KM/WM system centers on the ability of marketing management to anticipate changing external and internal factors over time. The manner in which marketing management relates to changing factors is one way of determining how well it is prepared to meet the challenges of changing times. An integral part of a company's marketing strategy is to put the customer first by creating and distributing unique values for them along with the company and its trading partners. In effect, the company's customers are linked in a partnership such that there is a win-win situation. The capability of marketing management centers on linking marketing strategy with its sales plans of today and tomorrow so that services to customers are maximized over time. Needless to say, leadership by marketing management is required to accomplish market strategies and plans within an optimal KM/WM system-operating mode. Emphasis is placed on the ability of marketing management in getting its staff to achieve specific sales goals over time. In the process, there must be a high degree of confidence and trust among marketing managers and their sales personnel. An optimal KM/WM system environment stresses the need for evaluation of the marketing strategy and sales plans as set forth by marketing management. This includes how well marketing goals are related to actual sales so that corrective action can be undertaken if results are below expectations. Results may indicate that the organization's business model in terms of tie-in with a company's overall marketing strategy and sales plans must be changed for the times.

Relate Marketing Strategy and Sales Plans to Corporate Planning for the "Big Picture"

In this 21st Century, companies are spending millions of dollars on the development and maintenance of an appropriate overall marketing strategy and sales plans that are tied-in with its corporate plans that centers on the "big picture." Essentially, wisdom is being employed by decision makers at small- to large-sized corporations for getting an overview of where they are as well as competitors. Wisdom assists them in helping their customers to sustain their success. As noted in the text, wisdom comes from vast experiences over time that have been refined and reworked to improve current operations. For example, Procter & Gamble has undertaken such an approach to its marketing efforts such that they are more effective as well as efficient. It is testing the latest computer technology that will send a message to its database as soon as a customer lifts one of its products off the shelf. This package program has undergone trial runs. Whether or not the experiment succeeds, it is an important component in P&G's broader effort to retool is supply chain and kick start stagnating growth for a more effective marketing strategy. P&G is pursuing a newer and wise approach in its supply-chain management in order to manufacture products like Tide laundry detergent, Crest toothpaste, and Charmin toilet paper that are in direct proportion to actual consumer demand rather than consumer forecasts. It is a goal that companies in the consumer packaged-goods industry have been chasing for years. The supply chain — the process that takes a product from raw materials to manufacturing to distribution — can account for as much as 75% of a product's costs. Within five years, Procter & Gamble plans to go a step further by reducing its current 131 factories to 100.

To assist P&G decision makers in reaching its long-term marketing strategy and related sales plans, it is counting on futuristic technology. The computer packaging program, which uses technology developed at the MIT Auto-ID Center, is still in the very early stage, and it is unclear whether it will work or how widely it will be used. It may simply be a tool employed in a few locations to test consumer response to new products. Nonetheless, the idea of wiring store shelves is a novel one, and it could radically change the way the company tracks sales. The technology employed is such that small chips are attached to the packages of P&G products. When a consumer picks up, say, a tube of Crest toothpaste, a signal is sent to the store shelf, which contains printed circuit boards. The shelf transmits each "transaction" to a computer, which tracks the number of times consumers buy P&G products. That data is then transferred to P&G, which can adjust its manufacturing and distribution plans according to which products are moving off the shelves. Overall, the idea, as with the company's entire supply-chain effort, is to get sales data earlier and more often so there is a further improvement of the company's profits over time (Dalton, 2001).

Take an Enlarged View of Wisdom for Market Leadership

When developing a marketing strategy and related sales plans for an optimal KM/WM system, the area of market share of a company's products and services versus its competitors always comes up. The focus centers on "Who's best." Non-market leaders look at the market leader with an eye toward copying what seems to work. They want what they do not have. For example, once zero inventory, lean manufacturing, innovations around supply-chain management, and associated technologies such as scheduling and logistics software become the standard in the auto industry, firms in other industries sought to implement their own supply chains. They are looking for leaders to imitate, and any innovators in the supply chain space would be foolish to ignore who their customers (and potential customers) will or will not be in the future. The question those companies are asking is not which companies have the best supply chain management capabilities or whom to emulate. But rather, the focus should be on the best means for differentiating oneself from competition as opposed to merely conforming to industry standards. The end result of this analysis is an intriguing matrix of innovation and imitation. Innovators need to study the imitation patterns and pathologies of their competition in both innovative and imitative ways.

This approach to becoming the market leader (or one of them in the future) is an important means for a company to increase "share-of-market" (SOM). However, this conventional wisdom approach needs to be supplemented by one that centers on "share-of-customer" (SOC) (refer to the discussion earlier in the chapter). Instead of concentrating on one product at a time and trying to sell it to as many customers as possible, an enlarged view of wisdom centers on share-of-customer that concentrates on one customer at a time, that is, trying to sell that customer as many products as possible over the customer's lifetime. Basically, a SOC marketing strategy uses the Internet that allows companies to interact on-line cost efficiently with individual customers. Simultaneously, the increased capabilities for computers to analyze and apply that data more quickly, relative to individual customers, has given decision makers the ability to make rational decisions with respect to how one customer should be treated differently from another. Overall, a share-of-customer marketing was a business strategy waiting for the right computer technologies to be available before it could take off. Share of customer is a comprehensive understanding of individual customers, then meeting their needs across wider and continually expanding span of products and services. It is not a store in search of customers, but a customer base in search of products (Peppers & Rogers, 2003).

Taking the above-enlarged view of wisdom, that is, share-of-customer, a step further, reference can be made to online companies. A most critical factor is timing. Early movers who capitalize on meeting individual customer needs tend to grow faster than online companies who enter the marketplace at a later date. Early movers generate time- and size-related benefits such as brand awareness, scale economics, benefits from network externalities, and switching costs. Continued opportunities exist for innovators that recognize where technology adds value and reduces cost in the distribution chain. The bottom line is that online companies need to apply an enlarged view of wisdom and always be prepared to retool their approaches to innovation as the environment changes.

From another view, to assist marketing decision makers in leading their companies, a number of predictive analytics have been successfully implemented to support customer recommendations, customer value, churn management, campaign optimization, and fraud detection. On the product side, success stories in (1) demand planning, (2) just in time inventory, and (3) market basket optimization are a staple of predictive analytics. Predictive analytics are used to get to know the customer, predict customer behavior, better understand product demand, and forecast market dynamics. Overall, predictive analytics are helpful in leading a company's customer in new directions not found in the past (Agosta, 2004).

Connect Marketing Strategy to "Points of Wisdom" for Judging Soundly

An integral part of the above share-of-customer approach and adding more unique values to customers as times change is the whole concept of making a company's customers its partners so that "points of wisdom" can be connected for judging soundly about a company's operations as well as its customers. Essentially, a share-of-customer focus requires intimate and ongoing collaboration among all trading partners. This can be accomplished by demonstrating to all partners the benefits of exchanging information, knowledge, intelligence, and optimization results. *Collaboration* is an investment of a company's and its customer's time and effort. Once that effort is put forth, it takes time to replicate this level of intimacy with others. The bottom line is that collaboration can become an important barrier to entry by competitors.

In years to come, companies will have so refined their collaborative processes such that customers will not think of choosing competitors. For example, a financial services company could know almost everything about an individual customer's economic existence. Because the company will continually assist the customer in optimizing its finances, the customer will be willing to cooperate to the full extent capable. Partners and customers alike will share information, knowledge, intelligence, and optimization results and, in essence, collaborate because the benefits of such integration flow to all participants for optimizing their resources will be obvious. All in all, in a collaborative, share-of-customer driven economy, the focus is on trust. If the costs of dishonest dealings seem significant now, they will pale in comparison to what is at stake within an ecosystem of customers and partners built upon trust and mutual financial advantage.

To assist in connecting points of wisdom for judging soundly concerning a company's resources, reference can be made to Figure 3. As set forth previously in Chapter V on developing and implementing optimal KM/WM systems, a *cause-and-effect diagram* can be undertaken to determine where wisdom should be applied. In the area of developing appropriate products and services that add unique customer value — that takes into account a collaborative and share-of-customer approach — a cause-and-effect diagram can be drawn which relates specific products and services to each other as well as their relationship to the total picture. To accomplish this task, marketing decision makers in cooperation with market researchers need to utilize a venture analysis modeling approach. That is, each product or service needs to be assessed on its revenue, costs, and profits over its lifecycle. In turn, each product or service can be related to each other such that the root or basic product or service should be introduced first before the others are marketed. In this manner, products or services are not introduced before their time.

Now that a cause-and-effect diagram has assisted marketing decision makers in determining the logical flow of new products and services that add unique customer value, the next step is to connect points *of* wisdom that are essential to judging soundly about the entire organization. When referencing Figure 3, the important points of wisdom that need to be connected for effective employment of an organization's resources are: (1) develop appropriate products and services that add unique customer value and (2) evaluate products and services to determine whether they meet customer needs. The connect between these two important points of wisdom in Figure 3 allows a company's marketing decision makers to get essential information, knowledge, intelligence, and optimization results concerning what is really important to keeping a company's products and services relevant to the times.

To assist marketing decision makers in connecting other marketing points of wisdom, reference can be made to Figure 4. For the first basic marketing function — *marketing strategy and sales plans* — there are four detailed activities, the first of which was set forth above — develop better products and services that add unique customer value. This detailed activity can be linked to collaboration and is measured by the degree of loyalty and trust among the partners. The second detailed activity — improve market share using a newer approach to marketing efforts — relates to the effectiveness of marketing efforts that is reflective in market share increase or decrease. The third detailed activity — determine short- to long-range marketing strategies and programs — can be tied in with marketing objectives and goals and is measurable by marketing goals

Figure 3. The development and evaluation of a company's products and services that add unique customer value

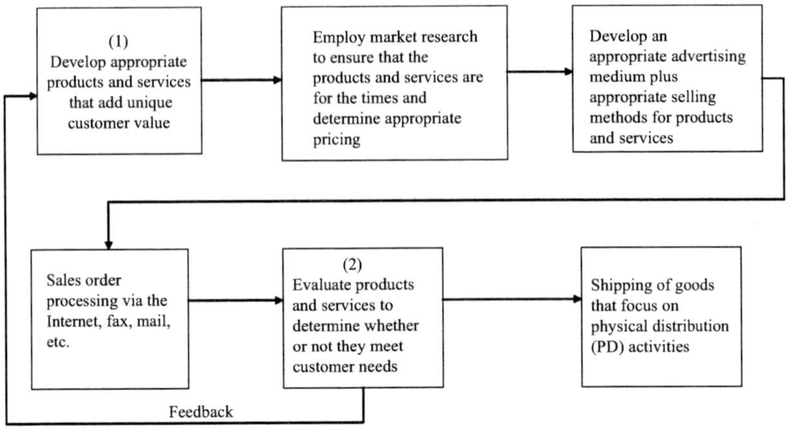

set forth in quantitative terms. The fourth and last detailed activity — reduce costs of products and services and pass on savings to customers — centers on the evaluation of products and services for the company and its partners in the form of cost/benefit analysis. In a similar manner, the other basic marketing functions — as set forth in Figure 4 — need to be expanded. They are: market research and pricing, advertising and selling, and sales order processing and customer service. These basic marketing functions can be related to detailed activities, which can be connected to points of wisdom making wise decisions. It is worth noting that the two major points of wisdom in Figure 3 are related to the first detailed activity for marketing strategy and sales plans and the second detailed activity for sales order processing and customer service found in Figure 4.

In summary, for a really effective and enlarged view to connect points of wisdom, it is quite helpful to have a collaborative and share-of-customer approach that allows customers, manufacturers, suppliers, wholesalers, and retailers to focus on logistics and the supply chain. In order for manufacturers, suppliers, wholesalers, and retailers to stay profitable is to understand their customers so that every transaction paves the way for future transactions by building customer loyalty. This ability to make fast and accurate predictions centers on the needs of the customers, thereby resulting in less inventory, lower prices, and loyal customers over time. The capability of connecting important points of wisdom is an important element in an optimal KM/WM system-operating mode (D'Antonio, 2003).

Figure 4. Relationship of a company's basic marketing functions and detailed activities to connecting "points of wisdom" for their measurement and analysis

Basic Marketing Functions (Figure 2)	Detailed Activities of Basic Marketing Functions	Connect "Points of Wisdom" For Optimal Decisions in Marketing	Measurement and Analysis Needed for Optimal Marketing Decisions
Marketing Strategy And Sales Plans	(1) Develop better products and services that add unique customer value	Collaboration among partners	Loyalty and trust among partners
	(2) Improve market share using a newer approach to marketing efforts	Effectiveness of marketing efforts	Market share increase or decrease
	(3) Determine short- to long-range marketing strategies and programs	Marketing objectives and goals	Measurable marketing goals
	(4) Reduce costs of products and services and pass on savings to customers	Evaluation of products and services	Cost/benefit analysis
Market research And Pricing	(1) Employ market research to ensure that products and services are relevant today	Correct reading of current needs for products and services	Social and economic analyses
	(2) Ensure that new products and services add value for customers	Additions of products and services for new markets	Cost/benefit analysis
	(3) Determine the appropriate prices for products and services	Pricing of products and services against competition	Pricing – part of venture analysis
	(4) Utilize optimization models to maximize a company's profits	Focus on a product lifecycle management approach	Venture analysis modeling

Figure 4. continued

Basic Marketing Functions (Figure 2)	Detailed Activities of Basic Marketing Functions	Connect "Points of Wisdom" For Optimal Decisions in Marketing	Measurement and Analysis Needed for Optimal Marketing Decisions
Advertising and Selling	(1) Develop an advertising media model in terms of costs and exposure rates	Examine different media approaches	Media cost analysis
	(2) Evaluate the advertising media model in terms of potential profits	Look for more cost effective advertising outlets	Advertising effectiveness analysis
	(3) Differentiate among the types of selling from one-to-one selling to mass marketing	Focus on newer forms of selling	Selling type analysis
	(4) Evaluate the need to expand selling efforts through innovative approaches	Reinvent methods of selling	Cost/benefit analysis
Sales Order Processing and Customer Service	(1) Processing of customer orders via the Internet, fax, mail, etc.	Classify by types of sales entry	Sales source analysis
	(2) Evaluate products and services to determine if they meet customer needs	Tie in original marketing efforts with current ones	Customer and profitability analysis
	(3) Tie in sales order processing with physical distribution (PD) activities	Explore physical distribution alternatives	PD service levels analysis
	(4) Evaluate the levels of customer service on a daily to a monthly basis	Periodic review of customer service levels	Customer service level analysis

Make Marketing Strategy an Integral Part of Venture Analysis Modeling

A company's marketing strategy can be represented in a number of venture analysis modeling applications. Typically, this means bringing together in one environment everyone and everything involved in product development, manufacturing, and service of the new product or service. VAM takes into account the fact that for most products or services, development is influenced by feedback from the marketing and field-support portions of the lifecycle. Venture analysis modeling interprets this extended chain as one whole process, rather than seeing it as a series of separate processes. For the most part, all participants need direct Internet access to product-related information. This collaborative approach has attracted large-scale adherents in the vendor community, including EDS, IBM, and SAP AG. Overall, a venture analysis modeling approach, which can be used for a thorough analysis of a product or service over its lifecycle, employs knowledge and intelligence about specific variables, assumptions, and constraints.

As shown in Figure 5 for a venture analysis modeling approach, a logical starting point is the external and internal factors sub-model ,which, in turn, is linked to the buyer behavior sub-model. In terms of the external and internal variables, all candidate variables need to be defined based on the judgments of marketing executives. Also, the competitive products and services must be defined in terms of how customers buy them as well as similar products and services. Next, a company product lifecycle management sub-model needs to be developed that is linked to the following: competitors' product or service sub-model, product or service characteristics sub-model, sales pricing revenue sub-model, variable and fixed costs sub-model, advertising and selling sub-model, and physical distribution and service sub-model. It is recommended that a data warehouse of historical marketing data, information, and knowledge covering not only a company's products and services be built, but also new and existing products by competition over the past several years for which accurate source data, information, and knowledge is available. It is from these figures that appropriate statistical and/or mathematical relationships can be developed for use by the appropriate sub-models. In turn, the output becomes the basis for short- to long-term profits, which determines the "go, no-go" decision regarding the feasibility of the new product or service.

Utilizing this venture analysis modeling approach for a new product or service, a marketing modeler can meet with a company's marketing decision makers who will supply various figures as they are called for by the various sub-models, including the estimated size of the targeted customers, recent product trial data, repeat purchases, the promotional budget, size of investment, target rate of return, product price, and gross profit margin for each product or service under study. The product lifecycle management sub-model will analyze the figures under computer control and display a forecast for price along with the total number of customers, company's market share, windfall profits (if applicable), period profits, and discounted cumulative profits. The marketing decision makers can alter various input estimates and readily ascertain the effect of the altered data on sales, costs, and profits. Additionally, prices can be varied over the life of the product or service. That is, the first year(s) allows a company to charge higher prices due to the newness of the product or service. As competition moves in, prices are generally lowered and profits tend to fall per unit. When the product or service is removed from the market or changed to meet changing customer needs, there can be a change in the pricing structure of the product or service to meet changed conditions in the venture analysis model. All in all, this comprehensive model assures marketing decision makers that the new product or service meets predetermined criteria, that is, revenues, costs, and profits — the last item noted in Figure 5.

Going beyond the venture analysis modeling approach (Figure 5), other types of software mentioned earlier in the chapter can assist marketing decision makers in cross selling and up selling at the point of sale. In addition, customized software

Figure 5. Venture analysis modeling approach that centers on revenues, costs, and profits over the life of a product or service within an optimal KM/ WM system environment

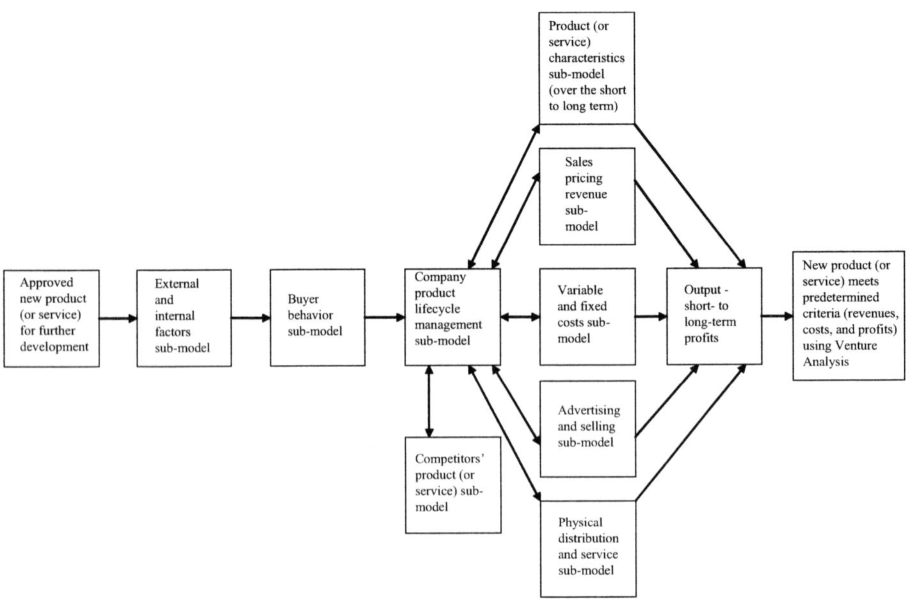

can be used to assist buyers and sellers in lowering their cost of doing business such that a larger view of wisdom can be undertaken. Many of the traditional functions performed by both buyers and sellers can be related to each other such that greater value is received by all in the process. Where they may be a question of who is to do what, that partner who is the most qualified in terms of adding the most value to the partnership should be given the nod. Where there is a good match, the acceptance and adoption of this approach should be evolutionary, not disruptive for best results for all parties involved. One that will produce clear business advantages, such as faster time to market, fewer design errors, and less cost, are the real measures of the effectiveness to making sound decisions for marketing decision makers.

Newer Insights into a Market Pricing Strategy

An integral part of a venture analysis modeling approach is the area of pricing. Conventional wisdom states that lowering the price of a product or service will allow more to be sold. However, in many cases, this price move is counterproductive since users of a company's product or service are looking for solutions, that is, newer ways to solve problems or pursue new opportunities. Hence, the value proposition comes into play and is at the very core of pricing. From another perspective, pricing can be related to the company's sales force. The traditional approach to selling focused on the sales force explaining the company's products and services and its pricing policies to its customers. A better approach centers on having all company employees who interact with the company's customers directly or indirectly explain more fully to customers the service value the company provides. This enhanced perspective in the eyes of the company's customers provides a means for demonstrating that pricing is but one element in the whole service stream. Essentially, this approach means focusing on an enlarged view of the customer and satisfying the customer's needs as they arise.

Because new insights, especially in the area of pricing, are needed for a robust marketing strategy, several vendors (other than noted earlier in the chapter) are focusing on setting prices that are right for current products and services. In the past, there were three basic ways to setting prices: cost-plus, market pricing, and out-of-thin air or gut level feeling. However, today, this *unscientific* approach has given way to a *scientific* one. Basically, scientific pricing is being shaped by a new set of insights into business strategy and human behavior, and these insights are integrated with software, mathematics, and rapid experimentation. Currently, there are several companies that offer scientific pricing services. For example, ProfitLogic (Cambridge, Massachusetts) works somewhat like airline pricing software to automate the markdown process. The computer absorbs several years' worth of data, looks at what is in stores, and how it is selling. In turn, it makes recommendations for prices on specific items. Casual Male picked ProfitLogic, which is also working with Home Depot, JC Penney, and Old Navy. In the first year, Casual Male did a test across six departments in all of its stores. Its buyers would tell the ProfitLogic software what inventory they wanted to move, what the price was, and, most importantly, when Casual Male wanted to be sold out or when the chain wanted to have a certain amount left to be sold at the outlet chain.

ProfitLogic's system not only gave guidance on what to discount, and by how much, but it also allowed Casual Male's buyers to ask their own questions, like, "What happens if we mark down 10% instead of 20%?" The bottom line was that the software indicated one basic change in the Casual Male markdown procedure, that is, discount less, but discount a lot sooner. Saks is using similar software, designed by Spotlight Solutions. Like at Casual Male, the system has recommended smaller markdowns, but sooner. The CIO (chief information officer) of Saks is well pleased with the accuracy of the software's algorithm predicting sales if prices are changed. The CIO believes that as the software becomes commonplace, it could have an unintended effect that of lowering list prices. In the past by overpricing at the beginning, it had to compensate for underpricing at the end of the season. Ultimately, everything may be a little cheaper, because sales will not have to absorb the cost of those deep markdowns at the end of the season (Fishman, 2003).

Going beyond a single firm, price optimization can lead to high profits for a supply chain. That is, e-sourcing can be used to gather a vast array of competing bids and to harness computing power to slice through the complexity of specifications and price variations, thereby allowing corporate buyers to reap unprecedented savings. Optimization can determine the ideal market-price range for a finished product based on the costs of goods and labor as well as determine the most favorable bids on all the parts and materials that go into the finished product. Price optimization techniques work best for businesses that need to buy a vast number of items with diverse specifications and a wide range of prices. Essentially, optimization lets members of the supply chain explore millions of solutions without having to look at each one individually. The technique can be most effective when a company is trying to allow the suppliers to differentiate themselves on things other than price whether it be warranty terms, payment terms, whether or not a minority supplier, or there is an existing contract. Currently, there are several vendors offering price optimization for a supply chain. One of the key benefits of optimization at this level is that it gives buyers a way to find real savings rather than shift costs between the buyer and seller (Pallatto, 2003).

An Optimal KM/WM System Application in Marketing

Optimal KM/WM system applications in marketing are making important inroads these days because marketing decision makers are typically more willing to try new approaches if the old ones fail. For example, at Wal-Mart, the whole concept of wisdom is embedded not only in marketing activities, but also permeates the entire culture of the company such that judging soundly is a way of life. Wal-Mart, which grew up in small-town America, is increasingly dependent on foreign expansion as it faces slowing growth at home. The U.S. market is somewhat saturated. Clearly, there is need to expand into foreign markets to generate the ongoing growth. The formula of low prices and good service works in other countries as it does in the United States. Wal-Mart grows the size of a Fortune 100 corporation each year. The company's culture is as strong as ever. If Wal-Mart were to maintain its average growth rate from the past 10 years, it would become the world's first $1 trillion company within a decade (Collins, 2003).

Within an optimal KM/WM system-operating mode, Wal-Mart has put together an efficient and effective supply-chain optimization system that allows it to connect points of wisdom in its everyday operations. (The subject matter of supply chain optimization will be covered in depth in Chapter IX.) Within this system, Wal-Mart relies on: (1) scale, (2) scope, and (3) speed. The economies of scale should be obvious. The scope part allows Wal-Mart to "flex" its toy section, as an example, before the holidays and collapse it afterward, while Toys 'R Us is stuck selling toys year-round. Scope also lets Wal-Mart use entire categories — gas, soft drinks, whatever — as loss leaders to pull people into its stores. The speed part is the most intimidating, that is, Wal-Mart's turnover is so rapid that 70% of its merchandise is rung up at the register before the company has paid for it. Also, speed is why many items never hit the warehouse floor, moving directly from truck to truck along miles of conveyor belts (Useem, 2003).

The bottom line for judging the soundness of Wal-Mart's decisions is adherence to a deeply democratic principle: Wal-Mart exists to enable people of average means the ability to buy more of the same things previously available only to people of means. The organization's business model of using its power to extract lower prices from suppliers and then passing those savings along to customers derives from that basic principle. Although Wal-Mart has been and willing to try all types of new ideas, the company keeps what works and gets rid of what does not. It always remains guided by its basic principle that is reinforced by sound judgment of what needs to be done organizationally over time.

Summary

To attain a higher level of marketing effectiveness within an optimal KM/WM systems environment, marketing decision makers need to undergo a major shift in their thinking about marketing operations that go beyond a conventional wisdom approach to marketing. But rather, the application of an enlarged view of wisdom to marketing centers on what needs to be done to grow the optimized organization over time. This exposition was followed by the factors to consider when developing a new marketing model for optimal KM/WM systems. They were: examine an enlarged view of market research and analysis, focus on a venture analysis modeling (VAM) approach, view customer relationship management (CRM) from a broad perspective, and employ marketing software useful in optimal KM/WM systems. A marketing model and its major sub-models were also examined. This discussion was related to developing products and services that a company knows what its customers want before they know themselves and influences the direction that the market will take today and tomorrow. In the next part of the chapter, a marketing strategy and related sales plans within an optimal KM/WM system environment were discussed. The tie in of conventional wisdom to an enlarged view of wisdom centered on a rethinking of marketing to assist in judging soundly about a company's marketing strategy and its related sales plans. Additionally, a case study of current optimal KM/WM systems was explored for marketing.

References

Agosta, L. (2004, August). The future of data mining – Predictive analytics. *DM Review*, 16-20 and 37.

Buzzell, R. D., & Gale, B. T. (1988). *The PIMS principles: Linking strategy to performance*. New York: The Free Press.

Collins, J. (2003, June). Bigger, better, faster (What I saw at Wal-Mart And what it means for the future of your company). *Fast Company*, 74-78.

Dalton, G. (2001, April 2). If these shelves could talk. *The Industry Standard*, 49-51.

D'Antonio, M. (2003, May/June). Merchandise optimization: Giving customers what they want. *1to1 Magazine*, 55-57.

Fayyad, U. (2003, May 13). Optimizing customer insight. *Intelligent Enterprise*, 22-26.

Fishman, C. (2003, March). Which price is right? *Fast Company*, 92-102.

Goldenberg, B. (2005, July). Real-time CRM: A business revolution in the making. *Customer Relationships Management*, 20.

Gupta, S., & Lehmann, D. R. (2002, May). What are your customers worth? *Optimize*, 63-68.

Keller, K. L., Sternthal, B., & Tybout, A. (2002, September). Three questions you need to ask about your brand. *Harvard Business Review*, 80-86.

Malone, R. (2003, January). It's the customer. *Managing Automation*, 40.

Moon, Y. (2005, May). Break free from the product life cycle. *Harvard Business Review*, 87-94.

Pallatto, J. (2003, February). Price optimization slices through data and wrings out savings. *Internet World*, 12-13.

Peppers, D., & Rogers, M. (2003, January/February). Prediction: "Share-of-customer" will win out over "share-of-market". *1to1 Magazine*, 56.

Peppers, D., & Rogers, M. (2004, October). Getting ready for your customer close-ups. *1to1 Magazine*, 50.

Schultz, D. (2003, April). You don't own the customer: The customer owns you. *1to1 Magazine,* 44.

Stackpole, B. (2002, August). What's the next step? Product lifecycle management. *Managing Automation*, 18-21.

Useem, J. (2003, March 3). One nation. Under Wal-Mart. *Fortune*, 65-78.

Vandenbosch, M., & Dawar, N. (2002, Summer). Beyond better products: Capturing value in customer interactions. *MIT Sloan Management Review*, 35-42.

Williams, P., & Stemper, M. (2002, May 6). Collaborative product commerce – The next frontier. *EBN*, 31-40.

Chapter VIII

Optimal KM/WM Systems in Finance

Issues

- To contrast conventional wisdom with an enlarged view of wisdom for financial activities
- To examine the important factors underlying the development of an effective finance model
- To explore the area of financial performance as it relates to an optimal KM/WM system
- To examine an optimal KM/WM system application that is evolving over time

Introduction

Typically, in the past, an organization's ability to manage and leverage its financial and related accounting operations went a long way to ensure its survival and growth. However, due to even faster changing times, an organization is particularly sensitive to its financial performance in the short to long run. A successful company in the 21st Century is one that does the best job of *leveraging its financial capital and intellectual capital*. It must be able to accommodate the massive changes in the business world, which include technological innovations, unexpected commercial alliances, industry transformations, and a host of powerful social forces. Many of these changes are linked directly or indirectly to the Internet. Astute financial decision makers need to employ the latest information systems technology that allows them to judge soundly about their decisions in the short run to the long run. In this manner, financial decision makers will be able to better optimize their financial and related accounting operations.

Within this context, the chapter examines initially a company's visioning and its challenging goals. The relationships of globalization to optimal financial decision making and the need to take a global financial and accounting viewpoint are examined. An enlarged view of connecting "points of wisdom" in finance that are set forth in a number of newer management principles. Next, the essential elements that underlie an effective finance model for optimal KM/WM systems are examined. In turn, a finance model and its sub-models found in optimal KM/WM systems are set forth. Because the text centers on making effective or wise financial decisions today and tomorrow, the area of financial performance is examined in some depth. In addition, an optimal KM/WM system in finance is given that takes a holistic approach to a firm that is experiencing growing financial problems. Overall, the main thrust of the chapter is assisting decision makers in accumulating financial wisdom to undertake the appropriate future opportunities and resolve pressing problems that cut through important relationships that connect "points of wisdom".

Visioning and Challenging Goals Underlie Effective Finance and Accounting Activities

Related to finance and accounting activities in an optimal KM/WM system-operating mode is a tie-in with a company's *visioning*. For a typical company, top-level decision makers must want their companies to be the leaders in their industries to better serve their customers. Visioning is not merely financial long-range planning intended to realize more sales, but a fundamental change in how the company must change to become more competitive or achieve a leadership role. Visioning is linked directly to realizing goals set forth by visionaries (who can be assisted by utilizing goal programming) and the development of new products and services (can be realized by using a venture analysis modeling approach). In turn, there is linkage to a company's critical success factors (CSFs) and key performance indicators (KPIs) along with financial ratios. Essentially, visioning centers on furthering a company's objectives and goals for growing an optimized organization. The ability of decision makers to judge the soundness of their financial operations about what needs to be done using optimal KM/WM systems is the underlying theme of this chapter.

Relationship of Globalization to Optimal Financial Decision Making

Throughout the text, the focus has been on the need for a paradigm shift — a move away from a conventional wisdom approach to a larger view of wisdom for optimal decision making. *Corporate planning* stresses the need to predict the future more accurately for what needs to be done organizationally. *Marketing* indicates the need for a fresh approach to products and services as a way of bringing the right ones to market for maximizing a company's marketing strategy. *Manufacturing*, as set forth in the next chapter, looks at the importance of organizational learning as well as bringing the customer into the supply chain so that desired values by all parties can be realized. In this chapter, *finance* will stress looking at newer financial methods and techniques that are more predictable than historical in nature for growing the organization. All in all, these ways of viewing wisdom "through new eyes" involve the process of reinvention in terms of how decision makers go about their jobs.

From an even larger perspective, financial decision makers need to look at globalization as a way to grow their organization financially over time. More specifically, all things being equal, if a company can build a product or render a service more cheaply by going overseas, the American consumer benefits as well as the company in terms of higher profits over time. Although globalization of manufacturing industries will slow down sometime in the future, this is not true in the near term. Typically today, companies of almost any size are being forced to increase their global scope in order to reap important financial benefits that add to a company's bottom line (Moad, 2004).

In the pursuit of globalization, a company's decision makers must engage in optimal financial decision making. That is, financial decision makers need to employ the latest optimization methods and techniques to increase ROI today by utilizing appropriate knowledge and intelligence learned from past and current financial analysis in order to predict the future more accurately. One such method would focus on the company earning more profit without more capital by employing optimization and cost-cutting methods in its manufacturing plants and warehouses that use the KISS (keep it simple, straightforward) principle. Another method would center on using less capital. For example, a beverage company could use plastic containers for concentrate instead of costlier metal ones. Still another method would focus on the company investing capital in high return projects. From another perspective, XBRL (extensible business reporting language), which is a recent global open standard for formatting financial information, makes reporting and analysis faster, cheaper and significantly easier to automate. It also improves access to information that often has been difficult to locate in financial datasets. In addition, XBRL reduces inconsistencies — such as ambiguous business terminology and myriad data formats — that often necessitate manual intervention and slow financial reporting. Overall, the employment of optimization methods and techniques plus XBRL are essential to the practice of optimal decision making that reaches around the globe today.

Need to Take a Global Financial and Accounting Viewpoint

In this era of globalization, there is a need to take a global financial and accounting viewpoint. In measuring and reporting financial results, companies in the United States have long followed the set of standards and guidelines known as generally accepted accounting principles (GAAP). But now that accounting scandals have exposed GAAP's vulnerability to creative interpretation, the Securities and Exchange Commission (SEC) and the Financial Accounting Standards Board (FASB) are opening the door to international influence on U.S. financial reporting. A shift toward global reporting standards is coming, and its impact on U.S. businesses will be far-reaching. Currently, there are a half dozen or more internationally important accounting standard setters, each with differing rules, including Australia, Canada, France, Germany, Japan, the United Kingdom, and the United States. But a harmonization of standards is now well under way. It is expected that in the near future that there will be only two accounting bodies with a dominating influence on global reporting: FASB and IASB (International Accounting Standards Board). Furthermore, it is anticipated that the two bodies will increasingly work to align their standards (Barker, 2003).

Additionally, financial decision makers should not forget the Sarbanes-Oxley Act of 2002 since it impacts all publicly traded companies, regardless of industry. This act is intended to bolster public confidence in the United States capital markets by imposing new duties and significant penalties for noncompliance on public companies and their executives, boards of directors, independent auditors, legal counsel, and securities analysts. The bottom line is that the Sarbanes-Oxley Act focuses on the need to manage and report accurate information.

Conventional Wisdom Regarding Management Principles Underlying Finance

A number of financial management principles have been developed over time that are based on conventional wisdom. A logical beginning financial principle is that right financial policies must precede financial tools. That is, it is essential to put the horse before the cart and not the other way around. It is necessary to complement a company's financial tools strategy with the appropriate organization and the right policies and procedures to make it all work together. Unfortunately, many companies seem to focus primarily on the tools factor, that is, hardware and software. Eventually, they come to realize that it takes a lot more than that. An effective financial system is based on optimization, governed by policies and procedures that are tied-in with the financial tools used by organization personnel for everyday operations.

An important financial principle is figuring out the best way to allocate a company's capital and which policies the company should follow to meet its growth objectives and financial goals. This may sound simple that the company makes more money than its capital costs and return the difference to its stockholders, but it is really not. Needless to say, choosing where to invest is an especially challenging job for a fast-growing company. Even in a stable industry, opportunities typically far outstrip a company's resources. The need to borrow seems to be on everyone's agenda, especially during times of expansion. The bottom line is a proper determination of allocating capital, whether it is generated internally or acquired externally.

Related to the preceding principle of the proper allocation of capital is one that centers on costs after the capital projects have been carried out. Essentially, the principle states that unexplained high costs in more than one area, such as marketing, capital investment, research and development, or any other area of a business can ruin any business. Too often, capital projects exceed their initial estimates by a small to a wide margin. This means that someone must make up the difference. Usually, this is where stockholders get hit if the cost overruns are substantial enough. Reading daily and weekly financial newspapers will verify this important point. This principle comes into play especially during periods of high inflation.

One principle that is often overlooked focuses on the *accountability* of the company's managers for financial performance. Accountability recognizes the significance of utilizing flexible budgeting and rethinking cost accounting approaches, among others, to improve the quality of information and knowledge provided for all management levels. Within such an approach to financial operations, finance takes on a new perspective. It goes beyond allowing managers to review current operations. It adds a new dimension by providing a means of synthesizing a myriad of information and knowledge for answering key managerial questions for one or more areas of accountability. Typically, such questions include: What happens to net profits before federal income taxes if variable costs of producing certain products are decreased by certain capital investments? What impact do interest rate changes have on net profits when funds must be borrowed from multiple sources on the outside if future economic conditions are truly uncertain? Although answering this type of question generally requires the use of mathematical and statistical modeling or the use of knowledge discovery software, an effective accountability approach can help supply answers that have far-reaching impact on the company.

In the past, the two main benefits of reengineering efforts were increased productivity and lower costs of operations. For the most part, reengineering in the past focused primarily on paring bloated organizations and eliminating unnecessary costs. This approach boosted profits in the short term but not in the long term. Hence, for short- to long-term growth in a typical company, a different financial principle must be applied which states that reengineering efforts are more successful when the focus is on improving the inflows of revenue. Without creating new revenue, reengineering efforts are not likely to produce increasing profits. Research has indicated that the market value of companies that emphasized growth expanded at a faster rate than the value of companies that focused on cost-cutting measures.

From a slightly different perspective, an underlying financial principle that stresses the importance of employing newer computer technologies allows a company to reduce its own costs of operations, thereby allowing it to reduce prices charged to its customers. Generally, such improvements are quite noticeable to customers, who are looking for improvements from this standpoint. This concern for customers goes a long way. It not only encourages them to buy more, but it also makes them partial recipients of the improvements that the company has made to its own operations over time. It may well be that the company's software and, possibly hardware, can be used to effect savings for its customers. For example, this concern for customers by the McKesson Drug Company is found in their operations today. Both McKesson and its customers are benefiting from its lower costs of processing orders. This is a clear case of a win-win situation.

Last, but certainly not least, is the principle of building bridges from finance and its related accounting activities to a company's other functional areas. Despite the traditional view that financial systems belong only in one area unto themselves, more and more companies are turning to finance and related accounting systems to bridge the gap among functional departments by providing updated information and supportive knowledge for reporting, analysis, and corporate planning. All parties from the company president to department managers to sales representatives to plant managers need to have their own view of finance and related accounting activities. Instead of operating at cross-purposes, all functional area managers need to coordinate all of their business functions — from production management, inventory control, work scheduling, and purchasing to accounts payable, sales-order entry, and financial analysis. Building internal bridges among a company's functional areas provides the basic framework for constructing computer networks that are linked to outsiders, whether they are customers, suppliers, banks, unions, or government agencies. Building effective bridges from finance and accounting to others within and outside the company results in improved financial performance and productivity.

An Enlarged View of Wisdom in Finance

The various management principles as set forth above represent current thinking about those business areas that affect a company's financial operations. Building upon these management principles, then, are a number of newer ones, some of which are set forth below and summarized in Figure 1. Companies can improve their financial operations by recognizing that the whole is worth more than the sum of its parts. The more a company can "bundle" its operations, the greater is the tendency to simplify the customer's life in today's fast changing environment where time and cost savings are of the essence. Hence, this important principle makes use of bundling where customer and company costs can be kept at a minimum. It is not enough to combine services that complement one another if the bundle does not give customers anything more than they could get by purchasing each time separately. The most successful bundles let customers replace several confusing operations with a single, simpler alternative. For example, a telecommunications company offers local, long-distance, and cellular phone service; paging; Internet access; and teleconferencing. But, unlike most of its rivals, it offers all those services in a single package, backed by a unified sales and service system, and tallied on one monthly bill. Because customers appreciate this convenience, customer retention rates have jumped greatly. Also, its charge-off rate on accounts has dropped correspondingly.

A most important newer financial principle centers on making sense out of a company's financial measures and applying the results to its everyday operations. A company's critical success factors are important measures that are developed by top management and monitored so that overall results are measured on a periodic basis. Going a step further, KPIs and financial ratios are the outcome of analyzing a company's balance sheet and income statement in some detail. Overall, a thorough, periodic evaluation of a company's financial operations can become the basis for corrective actions where deemed appropriate.

Related to the preceding principles is one that focuses on investing in people, that is, training and improving employee skills to improve a company's *return on knowledge* (ROK). Typically, the big reason that a company spends so much on training or electronic learning is that it is an investment in employee retention. Because most people today have seven jobs before their careers are over, what they look for in an employer is a continuous investment in their skills. Retaining employees is good for any company; however, this is especially vital for fast-growing companies.

Figure 1. Newer management principles help financial decision makers get an enlarged view for connecting "points of wisdom" about what needs to be done over time

- The more a company can "bundle" or consolidate its operations from a customer's perspective, there is a tendency to simplify a customer's life where time and cost savings are of the essence today.
- Evaluation of a company's financial operations via key performance indicators and financial ratios that tie back to a company's critical success factors forms the basis for corrective action.
- Investing in people, i.e., E-learning and improving employee skills, centers on a return on knowledge (ROK) which is just as important as obtainment of a high return on investment (ROI) for capital funds.
- Effective budgetary procedures should allow employees to measure themselves against the performance of competitors and against internal peer groups. In this manner, employees can focus their energies on those factors that are critical to beating competition.
- Keeping a company's finances under control needs to be addressed before the actual investment is made, during the actual implementation phases over time, and at the very end. In this manner, the soundness of financial investments can be assessed throughout the investment lifecycle phases.
- A sustainable and viable organization is one whose actions are designed to lead to a desirable future state for all stakeholders. From an overview position, a desirable future state includes maintaining viability and profitability as well as managing risk while promoting innovation.
- The chief financial officer (CFO) should be an excellent business advisor who has a good grasp of how the organization runs and makes its money. Also, the individual should have a full understanding of accounting, risk management, taxation, and all of the financial tools an organization needs to run and monitor its businesses.

Today, many companies are rejecting centralization as being too inflexible for financial planning. Typically, companies have invested large sums in computer networks, process reengineering, and a range of tools including balanced scorecards and activity-based accounting. But they have been unable to establish a new order because the budget and the command and control culture that it supports remain predominant. Conventional wisdom regarding budgets needs to be replaced by a more encompassing approach. Instead, companies need to allow employees the ability to measure themselves against the performance of competitors and against internal peer groups. Basically, employees do not know whether they have succeeded until they can appraise the results of a given period. Hence, they need to focus their energies on those factors that are critical to beating competition (Hope & Fraser, 2003).

In the past, very large corporations, like the Enron Corporation, have got themselves into trouble by not policing off-balance sheet financing properly. A truly wise company has the ability to get its hands on all aspects of income, expenses, and profits over the life of off-balance sheet investments. The guiding principle here is that such business deals should provide for standard financial measurement beforehand, during the actual implementation phase(s), and at the very end of the investment. As such, full disclosure is a necessity for all aspects of off-balance sheet operations. The bottom line is that the soundness of the financial investment can be assessed throughout the investment lifecycle phases.

An important factor today for companies and their investors centers on the sustainability of businesses. Research has indicated that companies who manage a wide range of sustainability factors are better able to create long-term value for its stakeholders. A *sustainable organization* is defined as one whose actions are designed to lead to a "desirable future state" for all stakeholders. For investors, a desirable future state would surely include sustained revenue growth over the long term. For the company employees, it would include workforce diversity. Regulators and the community at large value environmental stewardship and social responsibility. Consumers seek useful, reliable, price-efficient products and services. From an overview standpoint, a desirable future state includes maintaining viability and profitability as well as managing risk while promoting innovation. Companies that actively manage and respond to a wide range of sustainability indicators are better able to create value for all these stakeholders over the long term. The employment of wisdom by a company's decision makers will go a long way toward meeting the goals and objectives of all stakeholders. Companies that actively manage a wide range of sustainability indicators are better able to create long-term value for all stakeholders (Funk, 2003).

In order for a company to take an enlarged view of wisdom, it must be reinforced by the *chief financial officer* (CFO). Today, the CFO is not just the score-keeper who runs the numbers for the management team. Instead, the CFO has become a business adviser, that is, someone with a good grasp of how the business runs and makes money, who also has a full understanding of accounting, risk management, taxation, and all of the financial tools an organization needs to run and monitor its businesses. Added to the CFO's advisory role is continuous learning so that sound judgment regarding the many facets of a company's financial operations can be realized for a truly optimized organization (Meyers, 2002).

Development of an Effective Finance Model for Optimal KM/WM Systems

To develop an effective finance model for optimal KM/WM systems, there must be an effort by financial as well as accounting decision makers to employ optimization techniques that are based on reliable past figures and future realistic estimates. Although this approach to financial and accounting operations is very desirable today, there is still a long way to go for many organizations. As emphasized throughout the text, typical optimization techniques, such as goal programming and product lifecycle management, can be found in corporate planning, marketing, and manufacturing (see next chapter) although their usage in financial and related accounting operations is generally limited today. Tied in with these optimization techniques is *problem finding* that allows for a more thorough analysis and understanding of the problems and opportunities faced by an organization.

Related to the development of an effective finance model within a wisdom environment are the following: (1) utilize business partnering for optimizing financial performance, (2) develop appropriate cost methods to maximize profits over time, (3) track financial performance that focuses on profit growth, and (4) employ finance software useful in optimal KM/WM systems. These areas are discussed at length below. It should be noted that an effective finance model recognizes clearly that a company needs to maintain its organization's business model to focus on profitability rather than wait for revenue growth to solve the profitability problem over time.

Utilize Business Partnering for Optimizing Financial Performance

Today, optimizing financial performance that includes accounting activities means going beyond internal growth opportunities and includes joint ventures rather than just acquisitions. Revenue-building contributions have been extended to sophisticated revenue hedging and research planning models that have positioned these companies for even stronger top-line growth. There is a need to make decisions about which development projects to fund for revenue growth and how to structure product and service franchises, acquisition possibilities, and licensing arrangements. Partnering or joint ventures with others can be a successful venue for generating revenue growth and improved profits for an organization in the long run. The capability for partnering reaches way beyond traditional accounting and financial operations, thereby allowing a company to build revenues and improve profits without the need to merge. In effect, partnering makes such organizations as effective as a wholly owned subsidiary.

Companies that have formed innovative partnerships include IBM, Honeywell, Merck, Rockwell, and Square D as well as others. In addition, schools today are turning to banks, supermarkets, high-tech companies, manufacturers, and even Hollywood for the resources they need to implement technology, mentor students, and provide a host of enrichment services. These innovative partnerships emphasize shared common objectives, shared risk, creative financing, and business benefits. Typical shared common objectives include consistent prices, improved procurement process, cost reduction, order fulfillment tracking, coordinated billing, and sharing best practices. These shared common objectives should include the use of optimization techniques that results in the best financial performance of all companies involved.

An integral part of business partnering is doing its business and related finances and reports on the Web. Typically, e-commerce reporting requirements relate to the internal and external need for information and knowledge. The *internal* reporting need mainly involves financial, operational, and customer relationship management reports that help the company manage day-to-day operations. The marketing staff may also require analytic reports about customer behavior to help them recruit new business partners and customers. *External* reporting needs have two types: one to provide information and knowledge to business partners about the state of their business activities within the B2B (business-to-business) e-commerce environment, and another to provide consumers with information about price differences, specifications, and complementary products. Also, there is need to support two types of users: remote and internal. Remote users access B2B transactions through the Internet, and internal users' access that information through the company's intranet. All in all, business partnering requires various types of reports to meet their needs.

Develop Appropriate Cost Methods to Maximize Profits Over Time

For U.S. manufactures to control their production costs, they have been meeting this challenge with integrated manufacturing, just-in-time inventories, factory automation, robotics, and the like (refer to the next chapter). However even with these manufacturing advances, costing systems have not kept pace. Generally, today's product-costing data are wrong, often by extremely large margins. Without more accurate costing methods, the bottom line is a continued competitive crisis. Hence, there is need to rethink the current cost-accounting methods in this 21st Century. More specifically, companies need to take a hard look at activity-based costing and target costing. In turn, these basic costing approaches can be linked to direct costing which separates variable costs from fixed costs. Similarly, there is need to employ cost optimization techniques to better realize cost savings for a company. Activity-based costing and target costing plus cost optimization techniques are discussed below.

The focus of *activity-based costing* (ABC) centers on those costs needed to produce a product. A starting point is analyzing a company in order to determine all of its production and support activities. All costs are then assigned to activities. Next, activities are measured and linked with the products that consume the activities. The total cost of the finished product is an accumulation of the activities required to make the product. In addition to assigning the costs to the products that actually absorb the activities, ABC identifies cost drivers and isolates non-value-added activities. With this information, a company can establish priorities that focus on eliminating or reducing non-value-added activities. Basically, the ABC method is a restatement of costs that allow a company to identify specifically the activities that are generating costs so that management has a better idea of why budgeted numbers are being exceeded. If a financial analyst looks at the production department section of a profit and loss statement in both traditional cost accounting and ABC accounting reports, the numbers would be identical on the bottom line. However, they would be categorized differently. More recently, companies have introduced *time-driven ABC* into their processes, which provides transparent scalable methodology that is easy to implement and update. This newer approach to the ABC method allows cost analysts the ability to detect inefficient processes, unprofitable products and customers, and excess capacity (Kaplan & Anderson, 2004).

Today, the ABC method has expanded into *activity-based management* (ABM). Where ABC attempts to measure a product's true cost, ABM uses cost information to evaluate an entire operation. The goal is to distinguish between value-added costs (necessary) and non-value-added costs (unnecessary), and minimize the latter. As an example, take a factory that has trouble delivering on time. Many such plants have a group of employees called *expediters*. The expediters' job is to speed up certain products through the production process so that customers get them sooner than they would otherwise. ABC can capture all the costs associated with expediting — the people, the cost of obtaining missing parts, the freight to get goods to the customer quickly, and so forth. Managers using ABM, however, would observe that expediting in general is a non-value added activity. Rather than just counting up all of the costs, a manager would be smarter to make the production process faster and more predictable so that expediting is unnecessary. Although activity-based costing is quite useful to assist an organization in telling the real story about costs, there is a more pragmatic way to get a handle on costs, that is, take a Japanese approach.

Essentially, the Japanese look at costs before the fact rather than afterwards. That is, a Japanese cost-management system guides and motivates planners to design products at the lowest possible cost, and gives them considerable freedom in introducing new products as well as getting them to market quickly. In contrast, typical American companies developing a new product, for example, typically design it first and then calculate the cost. If it is too high, the product goes back to the drawing board, or the company settles for a smaller profit. On the other hand, the Japanese start with a *target cost* based on the price the market is most likely to accept. Then they direct designers and engineers to meet this target. The system also encourages managers to worry less about a product's cost than about the role it could play in gaining market share. This strategic thinking approach is a big reason why the Japanese so often come out with winning products.

The main focus of the Japanese cost-management system is getting costs out of the product during the planning and design stage. That is the point at which virtually all subsequent costs are determined, from manufacturing to what customers will have to spend on maintenance. This target-cost technique, which is used by such companies as NEC, Sharp, Nissan, and Toyota comes in countless variations. The stripped-down version has several important features. The team in charge of bringing a new product idea to market determines the price at which the product is most likely to appeal to potential buyers. From this crucial judgment, all else follows. After deducting the desired profit margin from the forecasted sales price, the planners develop estimates for each of the elements that make up a product's costs: design and engineering, manufacturing, plus sales and marketing. Each of these is further subdivided to identify and estimate the cost of each component that goes into the finished product. The bottom line is that a target costing approach takes a hard look at costs before the fact rather than after the fact.

In addition to the above cost-accounting methodologies, management needs to implement *cost optimization techniques* to better realize cost savings for a typical company. Essentially, cost optimization techniques (like other optimization techniques discussed throughout the text) focus on the employment of mathematical and statistical techniques to solve complex cost problems whose results depend on hundreds, thousands, and even millions of interconnected variables. For example, cost analysts could use optimization to determine how many manufacturing plants are needed, where they should be, and what each of those plants should manufacture which products to minimize costs. Similarly, optimization can be used to consolidate local and regional processing centers from a large number to a smaller number in order to centralize certain tasks and make their cost structures more scalable with current demand. Or to take a transportation example, cost analysts might use optimization to create the best possible crew schedule that takes into consideration routes that need to be manned. In all cases, the focus is on reducing costs in order to improve a company's bottom line.

Track Financial Performance that Focuses on Profit Growth

An organization's starting point is tying its current business model to appropriate critical success factors and, in turn, to focus on profit growth. Typically in the past, companies with the largest market share would have the largest revenues and the lowest costs for the highest profits. However, many large companies, like General Electric, General Motors, and IBM, have found this not to be true. What is happening is that companies are seeing a migration from products and services to what might be called "spinouts." Hence, the newer business model for changing times is centering on spinouts as drivers of profitability for growing the organization. For example, General Electric's jet engine profits are not in building the engine, but in financing, service, spare parts, and overhauling existing engines. Essentially, manufacturing of jet engines has become almost incidental. The important message here is that a typical company must change its organization's business model to reflect changing times before appropriate critical success factors and their measurement can be made that are of value to management (Calvin, 2001).

A company's *critical success factors* (CSF) are essentially the limited number of areas in which results, if they are satisfactory, will help ensure successful performance. They are the important areas where things must go right if the company is to flourish. If results in these key areas are not adequate, the company's efforts for the period will be less than desired. Typically, CSFs are areas of activity that should receive constant and careful attention from management. Performance in these areas needs to be measured on a continuing basis and the results should be available to higher management levels. CSFs can be best facilitated by a company's culture where organization personnel believe what they are doing is worth doing on a day-to-day basis. Since CSFs need to be reviewed for relevance to the times, today's CSFs may not be appropriate for tomorrow.

Within a CSF framework, there is need for higher levels of management to employ problem finding in order to relate what is known and what is not known to what needs to be known for a company to have the capability to be one step ahead of its competition. The knowledge gained from this approach includes the discovery of new emerging patterns and trends and provides the basis for proposing new strategies that center on improving a company's competitiveness and its return on investment. Such an approach also provides new insights for further analysis. The bottom line is that a company's critical success factors, such as the impact of the Internet and the World Wide Web, need to be changed over time to reflect the current business model.

Related to CSFs are a number of *key performance indicators* (KPIs) that are useful in determining its viability over time. The alignment of KPIs with an organization's vision/mission/objectives/strategies is the key to realizing bottom-line results. The challenge is to develop KPIs that provide a holistic and balanced view of the business. Faced with potentially hundreds of candidate metrics, a good, practical approach is to think of individual KPIs not just as a single metric, but as a balanced metric that incorporates several alternative dimensions. These dimensions include business perspectives (i.e., marketing, financial, process, and development), measurement families (i.e., cost, productivity, and quality), and measurement categories (direct, additive, and composite). By overlapping these various dimensions, a framework for developing key performance indicators that captures a company's most critical business drivers can be created (Bauer, 2004).

KPIs in marketing typically include the following. The ability or inability to make changes to a company's marketing strategies in order to increase market share is a beginning consideration for KPIs. The ability of the company to increase customers' satisfaction is a most important factor. The ability of market research to bring new products and services to market over time is critical to the growth of the company. In turn, the acceptance of these new products and services by customers is paramount to the company's success. Related to these KPIs is the ability to increase quality of products and services as well as improve delivery times. In terms of sales expenses, KPIs are related to decreasing them. The net results are KPIs that relate to gross profit and cash flow for a company's products and services over time. A number of KPIs can be developed not only for a company's other functional areas, but also for the company as a whole, which tend to focus on financial ratios. Overall, since KPIs are those over which a company has major control, a typical company should be able to manage the performance of these areas and make changes when deemed necessary.

Tied in with KPIs are a number of *financial ratios* that are useful to financial analysts is return on investment. ROI analysis can be used to judge present performance and to evaluate future investment opportunities. Analysts can also use ROI to rate managerial effectiveness and to compare potential profitability of divisions and departments. While the technique of ROI analysis is uncomplicated, the potential applications are varied and valuable. Fundamentally, the ROI equation is earnings divided by total assets, then multiplied by 100 to give a percentage. A problem in calculating ROI can arise in identifying what is meant by earnings or total assets. Although there is no single, correct way to figure ROI, it is customary to use earnings from operations before taxes, sales after returns and allowances for bad debts, and net year-end book value of assets. It is important that there be consistency and the same measure of earnings, sales, and assets be used when figuring ROIs for different periods. Also, managers should be prepared for apparent surprises when comparing the company's ROI to another company's, that is, the other company may have used different measures in calculating its ROI.

Complimentary to ROI is economic-value added (EVA). For a company to be truly profitable, it must have money remaining after it deducts the cost of all the capital it employs, that is, both equity and debt. Until a business returns a profit that is greater than its cost of capital, it operates at a loss. Hence, what is called profits, the money left to service equity and debt, is usually not profit at all. Related to ROI and EVA are a number of financial ratios that can be applied to a company's operations. Typical ones (which measure some aspect of a company) are: capital turnover (management efficiency), return on sales (operation efficiency), cash flow (liquidity status), current ratio (liquidity status), investment status (measure of solvency), return on assets used (asset profitability), average collection period (receivable investment), inventory turnover (inventory utilization), undelivered commitments (days of sales in backlog), net worth debt ratio (credit strength), acid test ratio (immediate liquidation), and stockholders' earnings status (percent earnings available). Some of these financial ratios are explored in more detail later in the chapter.

Overall, these financial ratios focus on those aspects that assist decision makers in their financial strategic thinking. Financial ratios can be calculated as needed and thereby serve as the basis for management by perception as well as management by exception where appropriate ranges are assigned to each ratio. Typically, to calculate these financial ratios, there is need to interact with a corporate database or data warehouse. All in all, critical success factors, key performance indicators, and financial ratios are useful for financial decision makers within an optimal KM/WM management environment for judging the soundness of their decisions over time.

Employ Finance Software Useful in Optimal KM/WM Systems

The whole concept of corporate performance management as noted previously in Chapter VI is an integral part of corporate planning. Similarly, CPM can be considered an important part of finance and accounting since a company's functional areas are measured and evaluated in terms of meeting predefined financial objectives and goals. Tied to CPM is a wide range of finance and related accounting software packages that are designed to accommodate the Internet and the World Wide Web. Typical finance and accounting software found in optimal KM/WM systems include the following: (1) optimization, (2) goal programming, (3) alternative to budgeting, (4) financial analytics, and (5) balanced scorecard and performance dashboards. Each of these software approaches and their accompanying software is set forth below.

As demonstrated previously in the text, *optimization software* is quite useful within an optimal KM/WM system-operating mode. One of these software packages is an integrated Enterprise Profit Optimization (EPO) suite, offered by the Manugistics Group. It is a software solution that helps drive organizations to profitable growth through the simultaneous optimization of its supply-side and demand-side functions. Combining the features of supply chain management and pricing and revenue optimization, the Manugistics' EPO offering consists of two major modules: Profitable Order Management (POM) and Profitable Promotions Management (PPM). POM is designed to help a company more profitably manage and fulfill the customer order, from initial customer inquiry through delivery. Furthermore, POM enables a seller to provide improved customer service while simultaneously improving the probability of repeat business. Finally, the seller can better ensure that the goods the customer demands are available, can be delivered at the time requested, and at price that satisfies the customer and enhances margins for the enterprise.

Another example of an optimization technique is found in the revenue management package of Talus. With years of experience, it is now in the midst of "productizing" its expertise, offering software (albeit that still requires some serious customization from one client to another) that can help companies assess all the variables that go into pricing and, thus, profits. Because the software makes recommendations as opposed to automatically setting prices, it is useful in a number of industries. Basically, Talus works by gathering and analyzing a wide range of data, and incorporating human expertise. In the airline industry, for example, it would look at historical, year-to-year booking trends, recent trends, special circumstances, costs, capacities, and other factors, and recommend (or set) prices so that each flight leaves the ground having earned the maximum revenue possible. In automobiles, it might look at the cost of a rebate trying to find the best point between big, profit-killing incentives, and no rebate at all, which maximizes profit per unit but undoubtedly weakens sales. Within an optimal KM/WM system environment, the above financial and related accounting software packages are useful to decision makers in making optimal decisions about a company's resources about what needs to be done over time.

Goal programming software, as noted previously, is centered on handling a single objective with multiple subobjectives or multiple objectives with multiple subobjectives. Within this framework, higher-level objectives can be maximized or minimized first before lower-level objectives can be brought into the final solution. As such, a company's objectives and measurable goals can be connected mathematically so that a company' profitability can be pursued in an optimum manner over the short to long term. For example, if a company desires to obtain a weekly or monthly contribution to fixed costs and profits of a certain amount, goal programming could be used to determine whether this goal is feasible. If not feasible, "what if" analysis can be used to determine what adjustments must be undertaken to reach the desired weekly or monthly goal. Later in the chapter, goal programming will be integrated with production goals so that profitability can be maximized for the company. It should be noted that several major software vendors have included goal programming into their generalized problem solving techniques.

Today, there is a newer approach to budgeting, that is, an *alternative to budgeting*, which was set forth earlier in the chapter under newer management principles. For example, the software, Comshare's MPC, is a performance management application package that integrates forecasting, budgeting, analysis, and reporting functionality. Another software package with similar capabilities is available from Prophix Software. Due to the software's capability, financial managers are offered a wide range of what to see and analyze. Typically, during the monthly rolling forecast review process, managers and their financial analysts field numerous requests for cost of doing business analyses, operating analyses, and slicing and dicing of cost and revenue trends. Before presenting answers to those questions at the forecasting meetings, financial analysts can run through drill-downs of the cost and revenue data and information to date. They can sit down and look at some of the big items that may have changed compared to the prior year or to the current forecast. Right away they can identify which departments are contributing to those variance items. Also, they can identify all of the different pieces that are contributing to the variance items. Overall, this alternative budgeting software ties in with the discussion to follow later in the chapter on replacing budgets with a newer approach.

An important feature of current software packages is providing *financial analytics*. Essentially, analytics enable decision makers to gain meaningful financial intelligence and understanding from the mass of millions or billions of individual transactions stored in a typical database. Analytics applications typically begin by consolidating the transaction data into a separate database. Real-time tools are typically provided to help users analyze the information and knowledge, such as comparing profitability across different divisions or product lines. An aggressive analytics provider is the Oracle Corporation with its Strategic Enterprise Management suite. Other software includes Brio Metrics Builder and Blue Pumpkin Software Advisor and Activity Manager. Brio Metrics Builder offers business analysts a powerful platform not only for analyzing and quantifying business trends, but also for creating new performance monitoring tools from company data and distributing them to managers and users across the enterprise. Blue Pumpkin Software's Advisor and Activity Manager duo consolidate company wide data into viable, role-specific metrics that bind customer service objectives to the specific performance of the sales force structure.

Today, organizations are realizing that there is an overlap in the analytical processes of the organization. Recent developments in financial analytics have been made in these areas of "overlap." For example, by combining traditional financial measures (revenue and cost) with CRM information (customer history) and applying predictive modeling tools and techniques, companies can now project the future profitability associated with an individual customer or household. This can be referred to as customer value management or return on customers (mentioned earlier in the text) which can be linked to creating unique values for customers. Customer value management enables organizations to monitor continuously each customer's value to the business and act accordingly. Overall, mangers need to know what value may be lost or gained before making decisions about nurturing specific customer relationships.

An important software package today that was mentioned previously in Chapter VI on corporate planning is the *balanced scorecard* (BSC) *approach*. This approach can be looked upon as a control panel of an airplane by keeping track of a company's financial progress as well as its softer measurements. It should be noted that the balanced scorecard was first developed and used by the Nolan Norton Institute, the research arm of KPMG. It uses several evaluation factors that are tightly aligned to the corporate mission. While this method does use financial comparisons for scoring competing projects, it also includes customer satisfaction, retention, and market and account share information. It considers internal quality issues, like response times, costs, and new product introductions. Also, the model calls for an examination of the learning and growth factors associated with the proposed project, including employee satisfaction and information system availability.

A balanced scorecard approach for some decision makers sounds like executive information systems discussed years ago. However, a BSC approach allows users to access the Web client, thereby enabling more people in organizations to access analytical information. A balanced scorecard ties in easily with an optimal KM/WM system since it incorporates the different technologies that have been around for years — data warehousing, OLAP, distributed networking, real-time operations, and middleware — to actually pull all the scores from all the individuals and then consolidate them. In effect, it represents a comprehensive and optimized set of metrics to gauge the health and competitiveness of a company. In turn, these measures provide a standard of comparison between a company and other companies in the same industry or otherwise (Williams, 2004). Today, several software vendors market the software. They include CorVu, Gentia, Hyperion Solutions, Oracle Corporation, PeopleSoft, SAP, and SAS Institute. These vendors run the gamut from offering highly aggregated data views with little additional functionality to providing high-level insight with drill-down data analysis and forecasting.

Related to a balanced scorecard approach are *performance dashboards*, which allow a company's decision makers to monitor areas for which they are responsible. Key Performance Indicators, for example, could be displayed in real time to assist decision makers in taking appropriate action. To many decision makers, KPIs measure the business health of the company and ensure that all company personnel at all levels are marching in step to the same company strategics, programs, objectives, and goals. They also provide the focal point for company-wide standardization, collaboration, and coordination. As such, decision makers can react to one or more KPIs and drill down to the appropriate level of detail for explaining why the deviations are above or below the expected. Among the dozens of such software packages in the marketplace are Indicative Service Director and SLA Manager from Indicative Software, and SilkCentral Performance Manager from Segue Software, which can manage performance in large, mixed enterprise environments. Both provide monitoring agents for a wide range of platforms and technologies such as server operating systems, networks, Web servers, and application servers. They offer flexibility in configuration performance metrics and defining service levels, thereby removing the need for a large number of disparate monitoring tools in the company's datacenter (Biggs, 2005).

Finance Model and Its Sub-Models Found in Optimal KM/WM Systems

When developing a finance (and accounting) model that is related to other functional areas of an organization, it is wise to start with a *corporate planning model*, as set forth in Chapter VI. By way of review, company resources need to be allocated in an optimum manner, thereby providing a basis for medium- and long-range financial plans (refer to Figure 2). Related to a company's medium-to long-range plans are its short-range plans that take into account the sources of funding for capital projects as well as the costing methods and the analysis of financial operations. Essentially, the *short-range planning — current year's performance measures sub-model* — provides a basis for determining the profitability or lack thereof for the company's day-to-day operations. The current year's performance measures can be used for comparing *actual* against *planned* performance. It is recommended that there always be a standing 12-month plan for a company's performance measures. A monthly comparison of actual amounts to anticipated performance measures results in exceptions, some of which are favorable and some that are not favorable. Those that exceed estimated amounts by 5% or more should be investigated while those within the expected range are generally left alone. Where deemed necessary, corrective action can be undertaken immediately.

Related to short-range planning is *sources of funding for capital projects sub-model* (refer to Figure 2). Once all relevant information and knowledge have been obtained about possible capital projects, detailed analysis is prepared for each project. This includes a listing of the relevant costs and savings associated with the acquisition and the disposal. Many of these capital projects are related to those financial resources needed to undertake the lifecycle management of a new product or service. Net cash flows are computed so that a return on investment (ROI) and net present value (NPV) can be generated. Financial decision makers, then, compare each project in terms of the risk factor. When all factors have been fully assessed, a final recommendation is made. Once the project has been accepted for implementation, it is necessary to determine the best method of financing. Generally, the project is referred to the treasurer whose task is to determine the best method of financing as well as the appropriate sources of funding.

The *costing methods sub-model* (Figure 2) has as its input the newer methods, such as the ABC method and target costing. These newer costing methods have the potential to be integrated with optimization techniques, including goal programming, linear programming, and the transportation model. This investigation is useful in projecting new products, manufacturing facilities, and personnel requirements that minimize a company's costs. The real focus of cost reduction is allowing a company to reduce its prices to customers as well as reducing overall organization costs. In some cases, cost reduction has the potential to add unique values desired by customers. As examples, the ABC method might indicate that cleanup costs can be eliminated since its serves no useful purpose in the manufacturing process of the company's customers. Also, senior personnel could act as mentors to eliminate or reduce training costs. Similarly, inspection of assembly operations may be better left to assembly personnel where production-line personnel are permitted to shut down production if there are quality problems. Evaluation of these areas by using activity-based costing may signal improvements for increasing profits. Overall,

Figure 2. Relationship of the finance (and accounting) model and its sub-models to the other financial models within an optimal KM/WM system environment

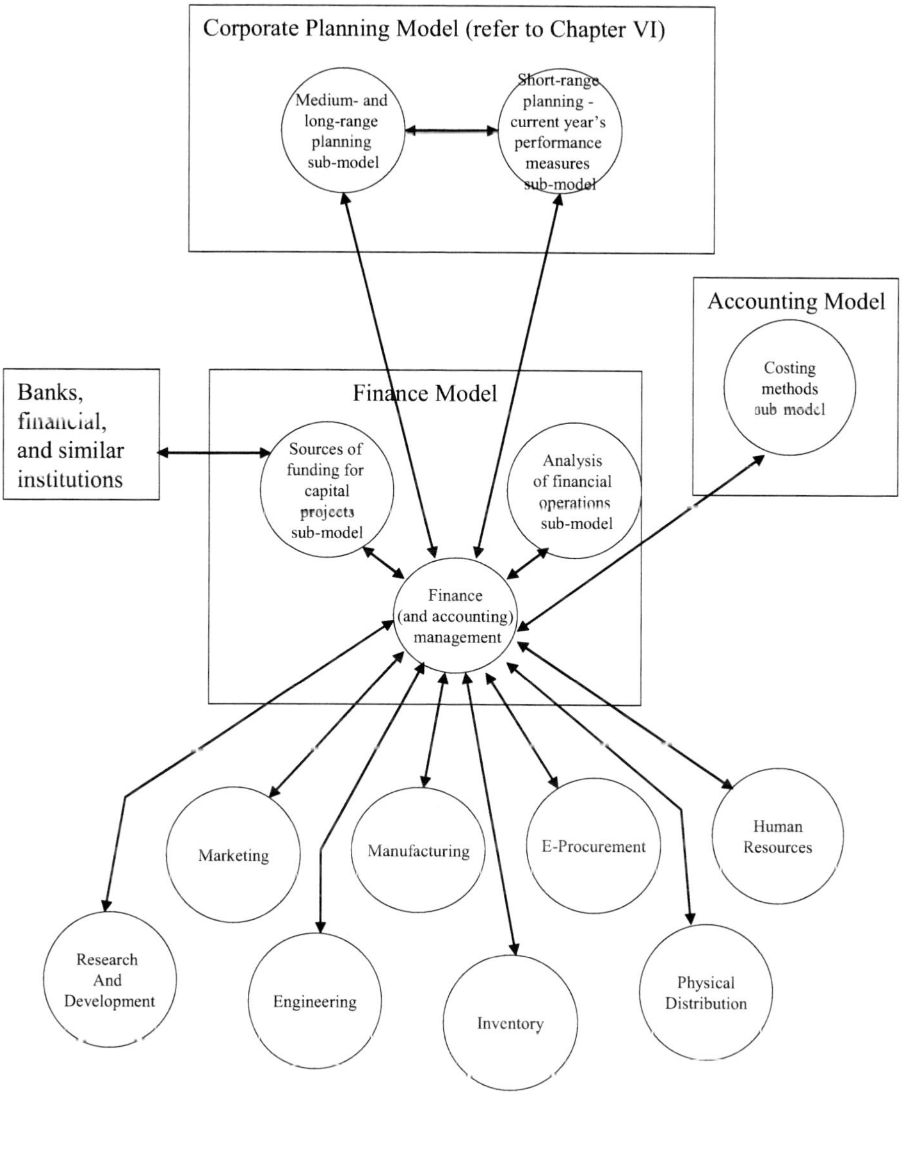

there may be need to realign plant facilities, the physical distribution system, reduce scrappage, and the like. From another view, there may be a need to think in terms of product lifecycle management. From an enlarged view of PLM, the venture analysis modeling approach accumulates the income and costs of activities over the entire lifecycle of a product or a service, from inception to abandonment by the manufacturer and the consumer. Its primary objective is a better match of revenues and expenses.

The last sub-model for finance per Figure 2 is the *analysis of financial operations sub-model.* Typically, financial operations are projected over the medium to long range, that is, projected balance sheets and income statements are developed two to five years and beyond plus estimated sources and application of funds. In turn, short-range financial plans for the coming year are developed, namely, monthly balance sheet and income statement plus sources and application of funds are prepared such that as one month passes, another month (one year hence) is added. In this manner, there is always a one-year standing plan. Actual comparisons are made not only for planned figures, but also to competing firms. In addition, comparisons are made among peer groups to determine who is doing the best job for furthering a company's financial objectives and goals. An extension of this sub-model is the capability of relating a company's CSFs to its KPIs and financial ratios. In all cases, these factors plus others need to take into account the optimization of a company's resources. Essentially, CSFs are tracked by key performance indicators and financial ratios. When KPIs and financial ratios conform to desired results (such as industry norms), there is no need for corrective action. On the other hand, when results are unfavorable, management at the appropriate level needs to initiate action. Generally, improving KPIs and financial ratios can be linked to the economic business cycle that confronts all organizations. However, this is not always true as can be seen in the failure of many organization business models. Hence, this last sub-model is designed to pinpoint financial areas that need improvement. Overall, the four sub-models found in Figure 2 for finance are useful within an optimal KM/WM system-operating mode since decision makers can better assess the company's financial soundness and make sound decisions about a company's operations today and tomorrow.

Financial Performance

For optimal KM/WM systems, a broadened view of a company's financial analysis and its related financial performance is based on information, knowledge, intelligence, and optimization of the company's operations today as well as tomorrow. From this broad viewpoint, financial performance for a typical company and its resulting evaluation is in the form of critical success factors, key performance indicators, and financial ratios. Initially, there is a tie-in with short-range corporate planning which forms the basis for evaluating current financial performance. Where budgets are not used, the use of rolling forecasts, which are created every few months and typically cover several quarters in the future, can be utilized for financial performance purposes. Because the forecasts are regularly revised, they allow companies the ability to adapt continuously to market conditions. It should be recognized that a financial performance system provides for individual managers to be assigned responsibility and, in turn, to be held accountable for meeting stated financial objectives.

Tie-In of Financial Performance with Corporate Planning for the "Big Picture"

Within an optimal KM/WM system environment, it is paramount that financial (and accounting) operations be integrated with corporate planning for getting the "big picture". A most important reason for reviewing a company's performance is to provide sound judgment to the company's top management and its corporate planning staff so that accurate financial projections can be made. By evaluating external factors as well as internal ones on a futuristic basis, the growth and profitability of the company versus its competitors can be compared by employing appropriate financial analytics, thereby providing a basis for an overview of operations internally as well as externally. For example, one competitor's projected future position can be compared to the company's own future financial picture. This projected analysis of the company's financial performance gives an indication whether or not competition is giving a fair return on its stockholders' investment. This overview of financial ratios needs to be supplemented by a more detailed analysis of competitors' financial statements frequently referred to as *content analysis* of present and projected financial statements and annual reports. Also, knowledge of forthcoming competing products and services that include their marketing philosophy is very important in the final analysis.

Basically, content analysis provides valuable clues to competitors' corporate strategy. Content analysis of competing companies along with their marketing policies and their future marketing directions can be useful for understanding specific issues of corporate strategy and can serve as a primary or supplementary source of information, knowledge, intelligence, and optimization results. It can be used to analyze current changes and past correlates of performance, and for more general investigations of potential future directions that are of interest to top-level executives and their corporate-planning staff. As an example, content analysis may disclose that one of the company's competitors is showing improved cost performance, that is, its costs are declining. The appropriate corporate response is for a typical company to get involved in a cost-analysis technique called *benchmarking*, which focuses on what the competitor does and how much it costs him to do it. In the company's lab, analysts tear apart the competitor's products and estimate the cost of designing and producing each part. The analysis extends beyond product costs. To pin down distribution and handling costs, company executives need to order some of its competitor's products, then trace where they were shipped from and examine how they were packed. Typically, cost savings on a particular product start at the earliest stage, with engineering determining the product design. The challenge is to find ways to make engineering more cost effective without stifling their creative efforts.

From another viewpoint on getting the big picture, some companies are moving away from flexible budgets (which may or may not reflect corporate strategy) since tight budgetary control processes tend to funnel information only to those with a "need to know". Companies that stay close to their customers so that they can respond quickly to intelligence about markets are rethinking budgeting — a process that disempowers the front line, discourages information sharing, and slows the response to market development until it is too late. A number of companies have recognized the full extent of the problems with budgeting. They have rejected the reliance on old data and the wrangling over what the data indicates about the future. In the absence of budgets, alternative goals and measures — some financial, such as cost-to-income ratios, and some nonfinancial, such as time to market — move to the forefront. Business units and personnel, now responsible for producing results, are no longer expected to meet predetermined, internally selected financial targets. But rather, they are judged on how well their financial performance compares with competition in terms of content analysis and benchmarks as noted above as well as with internal peer groups. Overall, a newer approach for assessing financial performance within an optimal KM/WM system-operating mode is moving away from the traditional budgeting process and focusing on what a company is doing versus its competitors. The ability of a company's financial decision makers to move ahead of competition in the future or, at least, keep up with competition, is a must that may require a different way of getting the big picture (Hope & Fraser, 2003).

Take an Enlarged View of Wisdom for Financial Performance

An integral part of taking an enlarged view of wisdom for financial performance centers on *problem finding*. Problem finding, when combined with an evaluation of financial performance, is quite helpful to financial decision makers to diagnose and solve many types of problems and related opportunities that are not possible with traditional problem-solving approaches. To illustrate the use of problem finding, consider a corporation, which has a measurable goal of increasing its value of its investments by 15% in five years. Top-level executives and its corporate planning staff, working in conjunction with the corporation's vice president of finance, the treasurer, and the controller, can analyze and solve the problem using different approaches. Each current investment can be evaluated thoroughly by examining ways to increase investment values. Likewise, each potential investment can be investigated using the product lifecycle management approach. If the current or potential investment's total return does not attain the corporation's goal, the important variables can be examined to determine what individual profit contributor(s) is (are) likely to fail to produce its share of gains. The net result is that these analyses are useful to identify current and future potential investment problems. In turn, the financial results are quickly calculated based on these problems, and the effect on the aggregate returns is seen immediately. Overall, problem finding can be quite helpful to financial decision makers to make sound decisions that affect the company's financial future.

Typically, conventional wisdom concerning the direction a company should take has centered on the growth objective, sometimes at double-digit rates. In turn, the financial performance of the company was measured based on the growth mentality. Many times, financial performance was less than desirable. Or, if financial performance was acceptable, there was no connection to a company's future financial performance. Thus, growth was limited in terms of its basic approach to optimize a company's profits over the life of its products and services. An alternative approach stressed centers around the value proposition such that all parties involved in the entire business process are the recipients of new and desired values not found in the past. This win-win proposition for all parties is paramount today. Hence, the need for open-mindedness about a company's structures, processes, and operations needed in an optimal KM/WM system environment should be on the minds of financial decision makers. Otherwise, the company could be headed in the wrong direction.

For example, in the roaring 90s of the last century, financial performance of many companies included some emphasis on financing the customer. Conventional wisdom seems to indicate that customer financing was a smart move when times are good. On the other hand, in somewhat bad times, the reverse appears to be true. The hangover of today of customer financing is by no means limited to victims of the Internet bubble's deflation. Boeing Capital Corporation, for example, is now caught in the downdraft of airline bankruptcies or near bankruptcies. Liquidity concerns confront the Ford Motor Company and Kmart-Sears, because of the huge short-term liabilities they have taken on to finance sales in the face of dropping demand. Hence, the message should be clear to companies — take an enlarged view of their operations before going further into customer financing. In fact, current wisdom may indicate the need to retrench in this area and pursue more promising opportunities that were not explored in the past (Fink, 2003).

One company has excelled at challenging conventional wisdom in the area of financial performance is Dell Computer. Basically, Dell makes additional profits from its suppliers by delaying paying its bills. In fact, Dell is one of the few companies that has a cash-conversion period of about 40 days. That is, it ends up receiving cash from its customers 40 days before it pays its own vendors. The company also is excellent at keeping its operating costs down; operating expenses, including research and development, as a percentage of revenue are about 10%, less than half of Gateway's. Dell Computer trades at a much higher multiple price/earnings ratio than its competitors. Though the company's stock price is not low, its above-average industry growth trends and the ability to turn over assets quickly make it a relatively good buy for investors (Schaff, 2003).

From another view of taking an enlarged view of wisdom for financial performance, progressive companies are turning falling prices into a chance at *reinvention*. Many industries have been suffering from sustained falling prices. Women's clothing and audio equipment, for example, are cheaper than in the past. Although this fact is good for thrifty consumers, unchecked deflation can cause havoc for a company, not to mention its employees and shareholders. It is no coincidence that some of the largest bankruptcies in history have occurred in industries facing massive pricing pressure. Deflation does not have to be a pain for companies. In fact, it can be a much-needed kick-start to innovation. Dell Computer has been living with deflation since its inception and which constantly reinvents itself to survive. Other large companies have gotten the message although their approaches are often very different (Mehta, 2003).

To assist financial decision makers in getting an enlarged view of wisdom for financial performance, virtual reality (VR) can be helpful. For example, a financial decision maker can transform financial information, knowledge, intelligence, and optimization results into colorful, interactive virtual worlds that can be explored in some depth. A financial decision maker can explore how money is spent at all levels of the company and can make better decisions on how the company's funds can be spent. In such cases, virtual reality software can create walkthrough scenarios of how a company can spend its money. Particular attention is given to flagging low-priority activities that were over-funded. Activities are assigned tall or short bars by their spending levels, and color-coded from green to red, to show priority. Exploring this 3D landscape, financial decision makers are able to determine which activities to view, touch, and manipulate. For example, they can focus on tall red bars, indicating low-priority activities with high costs. Touching these bars activates detailed information about the resources consumed. At any point, they could step back to see how resources were consumed for each high-level business process and for the entire company. Hence, virtual reality can be a very effective means to assist financial decision makers because it uses size, shape, and color in a 3D landscape to convey useful information, knowledge, intelligence, and optimal results that have been highlighted from extensive analytics. Similarly, virtual reality has the potential to turn a company's facts and figures about its financial operations into a giant simulated structure of the company and its competitors that is much easier to understand for making sound decisions that optimize a company's resources.

Connect Financial Performance to "Points of Wisdom" for Better Optimization

Related to taking an enlarged view of wisdom for financial performance is connecting "points of wisdom" for optimization of a company's operations and its resources. This enlarged view recognizes that an organization needs to assess which operations can die over time. However, well-managed companies are capable of outliving dying businesses. Typically, companies are reluctant to sell off underperforming businesses. The desire to hold on to businesses, in particular, successful ones of the past, is strong. This kind of sentimentality about businesses can undermine a company's continued existence. It is obvious that determining the right moment to sell is just as important as knowing the right time to buy.

Unlike companies, businesses appear to pass through predictable lifecycles, from robust early growth to maturity and on to decay and death. The length of those lifecycles can vary greatly, depending on, among other things, the stability of demand within a particular industry. Regardless of the industry in which they compete, however, some businesses will not reach their expected life spans, that is, they will fall victim to an internal disability or an inhospitable environment. Essentially, effective decision makers should do all in their power to help their business units thrive throughout their natural lives such that optimal financial performance can be realized.

In light of the concern for optimal financial performance being achieved by a company's decision makers, there is a need for connecting points of wisdom within the finance (and accounting) function. To assist in connecting the appropriate points of wisdom, reference can be made to Figure 3. As noted previously in the text, a company's critical success factors (CSFs) that are tied-in with *executive visioning* for a company's functional areas is an appropriate starting point for connecting points of wisdom for optimizing a company's performance over time. In turn, appropriate long-range, medium-range, and short-range corporate plans can be developed based on a company's CSFs. Specific sources of funding need to be provided for realizing corporate plans. In the process, pertinent costing methods are utilized so that funds can be accounted for in an effective and efficient manner. Appraisal of CSFs is undertaken by tracking key performance indicators and financial ratios as set forth in this chapter. The bottom line is that a company's CSFs should create long-term value for all stakeholders such that there is a win-win situation.

To better understand connecting points of wisdom in the area of finance (and accounting) as shown in Figure 3 that are related to a company's CSFs, there is need to appraise financial operations periodically. The goal of this appraisal is to ensure that the future financial performance will be optimum for the company so that wise financial decisions are being made by all levels of decision makers. A *cause-and-effect diagram* can be developed to determine where wisdom needs to be applied. In the area of CSFs, a cause-and-effect diagram that relates the various aspects of these factors critical to a company's success needs to be drawn in order to clarify one or more root CSFs. More specifically, this means looking at the following critical success factors: maintaining profitability that is above average, providing higher productivity at lower costs, providing values desired by customers, increasing customer loyalty, lowering prices while improving product quality, increasing profits over the life cycle of a product or service, responding faster to customer needs, getting products or services to the market faster, and similar CSFs. It should be noted that cause-and-effect diagramming procedures can lead to other factors outside the initial inquiry.

Having developed the appropriate cause-and-effect diagrams for getting at one or more root CSFs, the next step is to *connect points of wisdom* that are vital to the optimization of the organization. In reference to Figure 3, the major points of wisdom that need to be connected for effective optimization are: (1) specifying the critical success factors for the company's functional areas and (2) appraising CSFs by tracking their key performance indicators and financial ratios. The connection between these two basic elements in Figure 3 allows a company's decision makers to be fed information, knowledge, intelligence, and optimization results concerning what is truly important to a company's operations over the short to long run. In this manner, a company's decision makers have the ability to judge soundly and keep the company on course to meet changing times.

To assist in connecting other financial (and accounting) points of wisdom, reference can be made to Figure 4. For the first basic financial (and accounting) function, *short-range planning — current year's performance measures* — has four detailed activities, the first of which was set forth above regarding critical success factors. The remaining three are noted as follows. The second detailed activity — relate a company's long-range plans to medium-range and short-range plans to optimize performance — centers on a tie in of the long-range financial plans to medium-range financial plans which, in turn, are related to short-range financial plans. The third detailed activity — a financial performance approach supplements the budgeting process — looks at new performance measures that center on comparing competing firms as well as internal peer groups. The softer measurements of this newer approach can be found in the

Figure 3. The specification and appraisal of a company's critical success factors (CSFs) that are related to the finance (and accounting) function

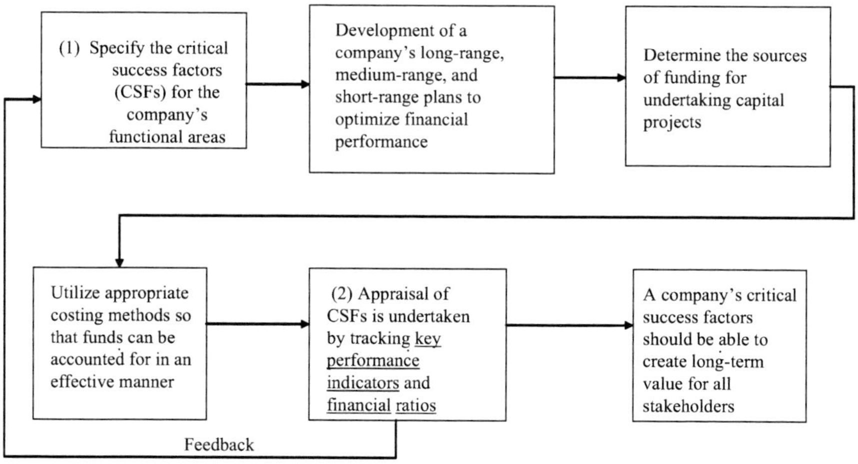

balanced scorecard approach. The fourth and final detailed activity — review financial performance with the view of improving future operations — has as its goal the capability of taking corrective action so that future financial performance will be optimum for the company. Similarly, the other basic financial (accounting) functions — sources of funding for capital projects, costing methods, and analysis of financial operations — in Figure 4 can be developed in terms of their detailed activities, which can be connected to points of wisdom for better optimizing a company's operations. It should be noted that the two major points of wisdom in Figure 3 are related to the first detailed activity for short-range planning — current year's performance measures and the first detailed activity for analysis of financial operations in Figure 4.

In summary, for a truly effective enlarged view to connect points of wisdom, there must be a never-ending drive toward increased profitability. For example, consider an insurance company that wants to target prospects that will become profitable customers over the long term. The most profitable prospects are those who respond, are approved, pay the premiums, stay for several years, and even buy additional products and services. To find such customers, it is necessary to define a "lifetime value" measure that consists of certain key components: approval, activation, and response. All of these components are combined in an equation that calculates a lifetime value measure. The values are derived from past experience and used to score and rank new lists of prospects. Essentially, lifetime value modeling can be entirely useful in helping organizations doing more with less, that is, improving profitability with less effort (Parr Rud, 2002).

Figure 4. Relationship of a company's basic financial (accounting) functions and detailed activities to connecting "points of wisdom" for their measurement and analysis

Basic Financial (Accounting) Functions (Figure 2)		Detailed Activities of Basic Financial (Accounting) Functions	Connect "Points of Wisdom" For Optimal Decisions in Finance (Accounting)	Measurement and Analysis Needed for Optimal Financial (Accounting) Decisions
Short-range Planning Current Year's Performance Measures	(1)	Start with specifying the critical success factors (CSFs) for the company's functional areas	Recognize the need for setting forth a company's critical success factors	Critical success factors analysis
	(2)	Relate a company's long-range plans to medium-range and short-range plans to optimize financial performance	Develop projected balance sheets, income statements, and source of funds	Short- to long-term financial statements
	(3)	A financial performance approach supplements the budgeting process	Measurement of competitors and internal peer groups	Content analysis and benchmarking
	(4)	Review financial performance with the view of improving future operations	Undertake corrective action to improve future results	Cost/benefit analysis
Sources of Funding for Capital Projects	(1)	Obtain all necessary facts about possible new capital projects	Examine all relevant income and cost factors	Net cash flow analysis
	(2)	Determine the risk factor associated with each new capital project	Examine the risk factor under good and bad times	Risk factor analysis
	(3)	Review the capital projects for meeting the company's established criteria	Project income/expenses for intended results	ROI and NPV (net present value)
	(4)	Determine the best method of financing the capital project	Relate external to internal financing for the capital project	Cost of capital analysis

Figure 4. continued

Basic Financial (Accounting) Functions (Figure 2)	Detailed Activities of Basic Financial (Accounting) Functions	Connect "Points of Wisdom" For Optimal Decisions in Finance (Accounting)	Measurement and Analysis Needed for Optimal Financial (Accounting) Decisions
Costing Methods	(1) Utilization of a number of newer costing approaches and methods	Classification of costs to better understand them	Cost/benefit analysis
	(2) Useful in projecting new products, manufacturing facilities, and personnel requirements for minimizing costs	Forecasting future requirements that reduce costs	Various forecasting techniques
	(3) Evaluation of the above areas for possible profit improvements	Relate current needs to future needs	Optimization techniques
	(4) Realignment of present facilities, an improved physical distribution (PD) system, reduced scrappage, etc. for lowering costs	Simulation of plants and PD systems to understand their operations	Cost reduction analysis
Analysis of Financial Operations	(1) Appraise critical success factors that create long-term value for stakeholders	CSFs are connected to KPIs and financial ratios	Critical success factors analysis
	(2) CSFs are tracked by key performance indicators and financial ratios	KPIs and financial ratios are specified for each area	KPIs and financial ratios analysis
	(3) Comparison of KPIs and financial ratios undertaken	Comparisons can be made continuously	Real-time analysis
	(4) Need for corrective action of unfavorable KPIs and financial ratios	Corrective action can range from minor to major	Cost/benefit analysis

Optimizing Financial Performance Requires the Application of Appropriate Analytics

To better optimize a company's financial performance for connecting points of wisdom, there is a need to apply appropriate financial analytics that are tied-in with an organization's big picture. Typically, this means starting with cash forecasts that are integrated into short-term corporate plans for the coming year. Also, included are capital budget requirements for the company's new products and services. As such, finance decision makers need an understanding of past and current operations in terms of sales, costs, and profits. More specifically, not only can total sales and related costs be calculated and illustrated for the coming year and beyond, but also the net profit before federal income taxes can be calculated and shown for the same time period. A data visualization tool can be used that depicts a graphical comparison of sales, variable costs, contributions, fixed costs, and net profits before federal income taxes (shown on the x axis) for a company's current year with the past four years (shown on the y axis) along with the amounts (shown on the z axis). The net result of comparing a company's current year with the past is that these data can be ranked from best to worst.

For an effective way to optimize financial performance, reference can be made to goal programming. For example, consider a company that manufactures two products in one of its plants. The goal set by its president is to maximize contribution to fixed costs and profits before federal income taxes of so many dollars per day. Based on this goal, this manufacturing plant has a current operating capacity of 500 hours per day. Production of either product — A or B — requires one hour in the plant. Because of current daily demand, only 300 units of product A and 400 units of product B can be sold, which must be factored into the problem. The contribution to fixed costs and profits from the sales of product A is $10.00 while the contribution from product B is $5.00. The single goal, then, of maximizing daily contribution can be broken into multiple subgoals or producing so many units of each of the products. At the same time, there must be consideration for the daily sales limits that cannot be exceeded. In light of the problem's dimensions, the problem can be defined in terms of its order of importance, which has been determined by the president. They are: (1) avoid underutilization of production capacity, (2) sell as many units as possible — since the contribution from the sale of product A is twice that of product B, the president is doubly anxious to achieve the sales goal for product A relative to product B, and (3) reduce overtime. It should be noted that this problem can be expanded greatly where there are multiple products with varying demands to reflect all aspects of the problem. Based on the constraints in the problem, it can be formulated and set forth in a first tableau (table) — refer to Table 1. Utilizing the procedures found in goal programming, the final values that would be obtained in the final table would indicate that all goals cannot be accomplished

Table 1. First tableau (table) of a goal programming problem that centers on maximizing a company's daily contribution for one of its manufacturing plants

		C_j	0	0	$-P_1$	$-2P_2$	$-P_2$	$-P_3$
C_j	Basis	Value	x_1	x_2	d_1-	d_2-	d_3-	d_1+
$-P_1$	d_1-	500	1	1	1	0	0	-1
$-2P_2$	d_2-	300	1	0	0	1	0	0
$-P_2$	d_3-	400	0	1	0	0	1	0
P_1	Z_j	-500	$-P_1$	$-P_1$	$-P_1$	0	0	P_1
	$C_j - Z_j$		P_1	P_1	0	0	0	0
P_2	Z_j	-1000	$-2P_2$	$-P_2$	0	$-2P_2$	$-P_2$	0
	$C_j - Z_j$		$2P_2$	P_2	0	0	0	0
P_3	Z_j	0	0	0	0	0	0	0
	$C_j - Z_j$		0	0	0	0	0	$-P_3$

completely since overtime cannot be reduced to zero. The optimal solution, then, is to produce 300 units of product A and 400 units of product B, with 200 hours of overtime in order to achieve the priority goals as nearly as possible. In terms of total daily contribution, the calculated value is \$5,000.00 (i.e., 300 units x \$10.00 + 400 units x \$5.00).

Another way to optimize financial performance is to examine a company's contribution. By increasing prices and decreasing variable costs as well as by decreasing prices and increasing variable costs will allow a company's decision makers to see the increasing or decreasing effect on contribution. Similarly, fixed costs can be reduced as well as increased to see the impact on net profits before federal income taxes. The profit trend line can be favorably influenced by increasing the sales price of the product, decreasing the variable costs, or improving a product mix. Change in the product mix can be helpful when several products are involved if increased concentration is obtained on products with high contribution. Changes of these factors in the opposite direction will naturally have an unfavorable influence on net profit before federal income taxes. This same type logic can be applied to fixed costs and net profit before federal income taxes.

Like with contribution, comparable financial information and knowledge and, in turn, appropriate KPIs and financial ratios can be developed. The intended purpose of these analyses by company financial managers is to determine whether the company is improving its financial stature over time. Especially important is the fact that financial information and knowledge discloses whether company managers are really managing effectively over the short run. Generally, this type of "at a glance" multidimensional analysis allows managers to start their thinking processes quickly. KPIs and financial ratios over a longer-time frame may indicate changing patterns and emerging trends that need to be explored by top management and their corporate planning staff. In some cases, analysis may indicate the need for problem finding to get the company back on track. The analysis may also signal the urgency of the company managers to change their financial understanding to a more global perspective that include the emerging nations of the world. From this view of changing world intelligence, it may be necessary for a company's top management and their staff to go back and ask questions about the company's critical success factors. Although current CSFs are the guiding force for top management, this may not be the case for tomorrow. Typically, financial results may indicate a change is necessary in the overall direction for a company that includes changing its CSFs. Accounting and financial managers need to look behind the figures in a typical balance sheet and income statement for optimal financial performance.

The balance sheet, which is a good snapshot of the company's financial structure, includes how funds (assets) are being used and from where the funds come (liabilities and equity) at a given moment coinciding with the income statement's end date. What the balance sheet measures is the financial efficiency of the company using the same industry data and standards applied to measure and compare performance. For most organizations, one of the most critical ratios to monitor is the *current ratio*, that is, current assets divided by current liabilities. This ratio measures the adequacy of the company's working capital and its ability to meet its short-term obligations. If the current ratio is less than one, then the company is clearly in trouble. Typically, industry standards are used to benchmark the most appropriate current ratio. For some industries, if may be 1.5, in others 2.0. A more stringent test of liquidity is the *acid test ratio*. Instead of dividing total current assets by total current liabilities, subtract inventory from current assets and then divide by current liabilities. For most industries, the standard is approximately 1.0.

Another key measure is the *debt-to-equity ratio*. This measures how the company's finances are structured. A ratio of three usually indicates that a company is using an appropriate mix of debt and equity to finance its operations. For most industries, a ratio of four indicates that it has reached its borrowing capacity without an additional equity injection. If the company has no equity, or even worse, negative equity, then its survival is highly unlikely. Thus, the debt-to-equity ratio is a vital measure concerning the viability of a typical company.

Related to the above financial analyses that connect points of wisdom in a typical organization is a *balanced scorecard approach*. This approach allows a company's decision makers to assign KPIs and provide them with the ability to track and optimize performance based on those indicators. KPIs are measured based on a set of metrics that consider multiple interdependent perspectives and they help organizations *balance* their focus on more than just the bottom line. This approach ensures that customer service, employee satisfaction, and sales and marketing are weighted appropriately, resulting in well-rounded and successful companies. For example, Hilton Hotels Corporation's balanced scorecard sits on an interactive Web-based portal that instantly is updated whenever new information is received. Not only do Hilton's managers have up-to-the-minute reports on their performance, the scorecard also provides "drill-down" information that helps them pinpoint the source of specific results so they can improve continuously. Essentially, a balanced scorecard approach for Hilton Hotels has taken center stage and transformed its hotel operations. It has allowed the hotel chain to implement, "What gets measured and managed, gets delivered." It is expected that this mode of operation will change to, "If you can judge it soundly, you will be able to optimize new found opportunities." A robust optimal KM/WM system platform for any type organization that uses a balanced scorecard approach plus a number of key performance indicators and financial ratios that are tied back to its critical success factors will prove to be successful over time.

An Optimal KM/WM System Application in Finance

Optimal KM/WM system applications in finance take a big picture approach to the financial performance of an organization. For example, the ratio of current assets to current liabilities should be about 2 to 1. Typically, to calculate these financial ratios, there is a need to interact with a corporate database. Basically, these ratios provide an insight into measuring actual results against the corporate plans (short-range to long-range). Management effectiveness can be measured by the capital turnover ratio, asset profitability, ROI, and the like. In addition, financial performance can be gauged through the return on sales, days of sales in backlog, inventory utilization, and receivable collections. An optimal KM/WM system application in finance, that is, Charles Schwab, that make extensive use of these hard financial measurements and its softer measures is set forth below.

A short time ago, Charles Schwab took the initiative to migrate its business to the World Wide Web, knowing that the move would force it to slash prices by about

60%. When the chief executive officer (CEO) was asked to describe the cause that imbued his colleagues with this revolutionary approach, he stated, "We are guardians of our customers' financial dreams." When was the last time a stockbroker looked like the guardian of a customer's financial dreams? Hence, Charles Schwab instilled the wisdom of making every employee feel that he or she is contributing something that will make a genuine difference in the lives of customers and colleagues. It should be noted that most people devote more of their waking hours to work than to home, family, community, and faith combined. Additionally, Charles Schwab credits its broad-based stock-ownership plan with keeping everyone focused on serving customers in ways that create new wealth. Since all employees own shares, everybody is on the same page. There is a recognition that company employees can get wealthy if they serve the customer better than someone else does. All in all, employee stock ownership plans are a beginning. Innovators at Charles Schwab need more than a stake in the company. They need a stake in their own ideas. There is a notion that rank correlates with wealth creation. To be productive, a company, like Charles Schwab, must disconnect compensation from rank, grade, job title, and hierarchy (Hamel, 2000).

To assist decision makers at Charles Schwab in carrying out their responsibilities, a balanced scorecard approach is helpful to them in assessing financial measures that are important to its customers and itself. The balanced scorecard approach allows Charles Schwab the ability to track its own financial results, which is needed to help nurture its growth today and tomorrow. In addition, this approach permits the company to assist its customers in improving their financial position over time. By the same token, there is the built-in capability of monitoring its competitors so that there is an assessment of what Charles Schwab is doing right and what needs to be corrected to maintain optimization of the company's resources as well as its customer over time. The capability of assessing hard financial measures along with softer ones is extremely important to Charles Schwab's top-level decision makers.

Like most brokerage firms, Charles Schwab has been adversely affected by the mutual funds scandal, bankruptcies of major corporations, low and negative returns on investments, the 9/11 attacks, plus other negative factors. After several thousand layoffs and a major restructuring, the company is working on new ways to grow its operations, like promoting its own branded mutual funds which are much more profitable than steering investors to somebody else's fund. Recently, however, after largely failing in its attempt to become a full service brokerage firm by luring the now affluent to its wealth management services, Charles Schwab slashed prices to appease its core discount customers and closed a number of local branches in an effort to cut costs. In effect, the company has moved back to its roots as a discount brokerage as much of Wall Street has shifted decisively in one direction or the other. Charles Schwab is utilizing its engrained culture as a basic framework for making a comeback (Cullen, 2004). Just recently, Charles Schwab is back as the chief executive officer of the company he founded. Although his succession plan did not work out, he is working hard to change the company for the times. He is utilizing appropriate management and computer skills to the fullest, which have the capability to help Charles Schwab sail through troubled waters not only today, but also tomorrow (Morris, 2005).

Summary

This chapter has focused on the application of optimal KM/WM systems to finance and its related accounting activities. The initial focus was on company visioning and its challenging goals, followed by the relationships of globalization to optimal financial decision making. Conventional wisdom and an enlarged view of connecting points of wisdom in finance were explored. Next, the underlying factors needed in the development of a new finance model for optimal KM/WM systems were examined in some detail. In turn, a finance model and its sub-models were presented within an optimal KM/WM system environment. This background served as a basis for a discussion on financial performance using optimal KM/WM systems. Financial decision makers were given new insights into their operations from a short-range to a long-term perspective. Lastly, a finance application was set forth. The bottom line is that financial operations and their processes will be performed in a manner normally not found in the past in order to optimize the profitability of an organization.

References

Barker, R. G. (2003, April). Global accounting is coming. *Harvard Business Review*, 24-25.

Bauer, K. (2004, October). Key performance indicators: The multiple dimensions. *DM Review*, 62-63.

Biggs, M. (2005, March 7). Dashboard for the enterprise. *InfoWorld*, 20-23.

Calvin, G. (2001, January 8). It's the business model, stupid! *Fortune*, 54.

Cullen, T. (2004, August 4). As Schwab returns to its discount roots small investors face shifting choices. *The Wall Street Journal*, D1.

Fink, R. (2003, March). What goes around. *CFO*, 44-52.

Funk, K. (2003, Winter). Sustainability and performance. *MIT Sloan Management Review*, 65-70.

Hamel, G. (2000, June 12). Reinvent your company. *Fortune*, 99-118.

Hope, J., & Fraser, R. (2003, February). Who needs budgets? *Harvard Business Review*, 108-115.

Kaplan, R. S., & Anderson, S. R. (2004, November). Time driven activity-based costing. *Harvard Business Review*, 131-138.

Mehta, S. N. (2003, May 12). How to thrive when prices fall. *Fortune*, 131-134.

Moad, J. (2004, October). Globalization reaches higher. *Managing Automation*, 18-23.

Morris, B. (2005, May 30). Charles Schwab's big challenge. *Fortune*, 89-99.

Myers, R. (2002, May). How CFOs stretch boundaries. *Journal of Accountancy*, 75-81.

Parr Rud, C. O. (2002, January/February). Lifetime-value modeling: The most valuable metric. *ltol Magazine*, 44-45.

Schaff, W. (2003, February 24). Dell succeeds where competitors lag. *Information Week*, 80.

Williams, S. (2004, October). Balanced scorecards in the business-centric BI architecture. *DM Review*, 34-39.

Chapter IX

Optimal KM/WM Systems in Manufacturing

Issues

- To examine conventional wisdom versus an enlarged view of connecting manufacturing "points of wisdom" for what needs to be done over time
- To set forth the important elements underlying an effective manufacturing model
- To explore areas that are related to production planning and execution
- To set forth a typical optimal KM/WM system application in manufacturing

Introduction

Over the years, a number of leading management experts have stated that employing newer technologies is the key to moving the country forward in terms of improved productivity. Prominent management experts maintain that newer technologies have not come close to yielding the returns that companies had expected, primarily because computer technologies and their related organizations are not properly aligned with business priorities or have not done enough to integrate internal business factors with external ones. Although each side offers valid points, newer information technology (IT) is indeed driving corporate productivity and GDP (gross domestic product) growth, but it is still only scratching the surface of its potential. Imagine the productivity that could be unleashed if technologists and management decision makers were all "working off the same page". This is the potential that optimal KM/WM systems offer to business organizations of all sizes, especially in the manufacturing sector.

In light of the above comments, the first part of the chapter looks at the need to rethink manufacturing operations from a broad perspective so that optimal decision making in this area is the order of the day. Next, conventional wisdom versus an enlarged view of connecting "points of wisdom" in manufacturing are examined. The requirements for developing an effective manufacturing model for an optimal KM/WM system are explored, followed by a manufacturing model and its sub-models in optimal KM/WM systems. Due to the importance of production planning and execution in determining what should be produced daily, it is examined in detail with particular emphasis on making it an integrated part of product lifecycle management. Lastly, an optimal KM/WM system application is given for manufacturing.

Need to Rethink Manufacturing from a Very Broad Perspective

In this 21st Century, there is need to rethink the entire manufacturing process from a much broader perspective in order to make it more efficient and effective. This rethinking centers on manufacturing concepts, particularly enterprise resource planning (ERP), supply chain management (SCM), manufacturing execution systems (MESs), and advanced planning and scheduling (APS). Additionally, the rethinking process should center on utilizing the e-learning concept, optimizing e-procurement and supply chain operations, utilizing total quality management (TQM) in the manufacturing process, and employing manufacturing software useful in optimal KM/WM systems. The major shift in thinking about manufacturing operations ties-in with the product lifecycle management approach noted throughout the text. Overall, a rethinking from a new perspective about manufacturing operations gives company decision makers and their staffs an improved way to optimize their operations that is not generally found in prior information systems.

Local and Global Manufacturing Requires Optimal Decision Making

Going beyond thinking in terms of local boundaries, manufacturing firms must continually reinvent themselves when going global in terms of their production facilities, which today also include the assistance of powerful supply chain technologies. Typically, customers want faster delivery, greater product customization, and order change flexibility, all of which are found in a collaborative universe. The ability to fulfill orders to demand without incurring excessive inventory and reduced profit margins is often the difference between success and failure. Optimal decisions must be based upon judging the needs of local and global manufacturing operations today as well as tomorrow. Typically, wise manufacturing decisions come from comparing alternative solutions after understanding each alternative in depth. Manufacturing firms must think locally in terms of optimal decision making, as well as connect globally for best results. An effective optimal KM/WM system-operating mode is not only good for local and global manufacturing, it is essential to grow manufacturing firms over time.

Over the years, manufacturing companies have found that one of the biggest issues with globalization is *communication*. It is not just language barriers, but communicating in the sense of coordination orders, inventory, production schedules, and quality testing. It is necessary to communicate in multiple time zones around the world with multiple local manufacturing companies as though everything is happening in real time (Neil, 2004). An integral part of this combined local/global approach to manufacturing is to facilitate product and process development over the Internet as well as sharing manufacturing capabilities. If the Internet has shown manufacturers and supplier anything, it is how business can be conducted with lightning speed. The question has now become: How can companies keep up? Judging by the zeal with which companies are expanding, the answer focuses on companies willing to think outside their local manufacturing operations and try new ways to run global manufacturing operations with trading partners.

Conventional Wisdom Regarding Management Principles Underlying Manufacturing

A number of management principles have been set forth over the years that are based on information, knowledge, and intelligence of manufacturing operations.

Underlying these principles for manufacturing is conventional wisdom, which is related to a thorough understanding of factory operations. To many on the manufacturing floor, these principles represent a common-sense approach to getting the job done. Essentially, the flow of information begins with customer requirements that, in turn, become the basis for detailed design. For example, a new vacuum cleaner is to be designed, marketed, and sold to the general populace. After consulting with representative customers and its retailers, marketing knows just how it wants the product to look. The engineer is concerned about the product's size and shape, not for appearance sake, but rather because the casing needs to house the circuitry inside that makes the product run. After the design process is complete, the computerized system also generates a bill of materials that will eventually feed into the company's manufacturing resource planning system. Based on this newer design and its processes, an important beginning management principle requires that all appropriate personnel within and outside the company participate and are involved in product development for a comprehensive approach.

For a company to maintain manufacturing dominance in its core products, an important management principle of manufacturing relates to the appropriate resources to reshape its product line to meet changing times. Control over core products allows a company to shape its evolution of applications and end markets. As a company multiplies the number of application areas for its core products, it must consistently reduce the cost, time, and risk in new product development. The bottom line is that well targeted core products can lead to economies of scale. Unlike physical assets, competencies do not deteriorate as they are applied but can grow over time to the company's competitive advantage.

Related to the foregoing is another important manufacturing principle. Since product lifecycles have been shortened dramatically in the recent past, companies have been forced to develop and commercialize new technologies faster than ever. As a result, competitive advantage now often goes to the company that is most adept at choosing among the large number of technologies and not necessarily to the company that creates them. This change in thinking by U.S. companies helps explain the resurgence of the U.S. electronics industry. It should be noted that a company's approach to this technology integration should suit the way it operates (i.e., its culture) and its resources.

In the past, many manufacturers have viewed their markets in terms of mass production. They have produced a great variety of goods and services. An alternative approach is to focus on customization. While mass marketing and mass production require that customers sort through a variety of options, customization offers them exactly what they want at a time, place, and price they want. By making the job easier for their customers, manufacturers gain allegiance for the long term. In essence, an effective marketing and manufacturing team finds and produces products and services for customers and not customers for its products and services.

Another manufacturing principle recognizes that manufacturing products in long production runs allows a company to move large quantities in as few shipments as possible. Today, there is agreement that companies must move products faster, fill orders more accurately, and do it at less cost. Reducing manufacturing cycle times enables better planning by more closely aligning production with demand. In some industries, such as apparel and high technology, market demands change every quarter, which means the need for reduced inventories.

An important management principle in manufacturing centers on cheaper and faster inventory turns. Turning over inventory faster is related to *just-in-time* inventories, which are designed to deliver raw materials and finished parts to where they are needed as close as possible to when they are needed. For example, car manufacturers operate on inventory that is delivered on a just-in-time basis, nor more than 2 to 4 hours before it is needed. Thus, the manufacturer's investment in inventory is cut dramatically today compared with just a few years ago. This has been made possible by giving suppliers access to sales data in a total supply chain management approach.

In addition, the key to reducing total inventory costs further is to recognize that for every dollar cut from inventory, companies save an additional dollar in warehousing, transportation, and other carrying costs each year. Needless to say, these cost savings can be substantial for any manufacturer. It may be possible to cut costs related to safety stock or buffer inventories. In terms of the total supply chain, all parties can benefit from a reduction in inventory costs.

Last, but not least, using a company's current manufacturing database in a productive manner is a most important challenge facing manufacturing decision makers. Because closing gaps is primarily achieved by connecting knowledgeable people from both inside and outside the company, manufacturing decision makers need to draw upon their established networking skills and use their traditional cross-functional roles to give them newer perspectives on effective manufacturing-based solutions. From this view, a company can utilize manufacturing information as well as from the company's other functional areas to meet its information needs.

An Enlarged View of Connecting "Points of Wisdom" in Manufacturing

An enlarged view of wisdom underlying manufacturing operations for what needs to be done over time goes beyond the management principles set forth above. As a starting point, it encompasses customers, suppliers, manufacturers, wholesalers, and retailers and the interaction among them in a total supply chain approach. To turn the *total supply chain* into a competitive weapon, product designers must initially be teamed with customers as well as personnel from their own marketing and manufacturing operations. Typically, a company finds it necessary to tear apart and rebuild its approach to product design. This means that designers spend time, using such electronic means as groupware, to get at all of the facts involving a proposed new product. A business team would normally consist of a representative from marketing, finance, and manufacturing and quality control to provide sound advice on the final product design.

Related to the total supply chain approach is *trust*. More specifically, this means that supply chain partners must have faith in the manufacturing company with whom they are dealing. In a similar manner, customers must have trust in the quality of the goods produced. Without credibility with all parties involved in the manufacture of the products, a manufacturer is subject to a wide range of potential problems. The lack of trust could result in the company's inability to leverage its multitude of technologies to boost its bottom line for not only itself, but also for all parties. In the end, the values promised for all parties could amount to nothing if there is a violation of trust, even on a small scale versus a large scale.

Related to trust is achieving *excellence in manufacturing*. This means making wise decisions when comparing different approaches and solutions in the area of manufacturing operations. Excellence implies accomplishment of objectives and goals and implies perfection. As applied to manufacturing operations, the notion of excellence embraces all of these ideas and more as companies seek ways, particularly in difficult times, to perform manufacturing operations better. At a basic level, achieving excellence means quality, reliability, and efficiency at all manufacturing levels. Drill down a bit and these ideas translate into reducing defects, increasing uptime, and streamlining operations, thereby yielding benefits such as cost savings and cycle time reduction. Underlying these benefits are important information, knowledge, intelligence, and optimization results from multiple points inside and outside the company for undertaking appropriate action, and sharing in real time to empower individuals, improve decision making, and respond to the needs of customers and the market with dispatch. The notion of manufacturing excellence gets more interesting when it is pointed in the direction of organizational agility and flexibility. For best results, manufacturing excellence is undertaken on a day-to-day basis.

In reference to agility and flexibility within an optimal operating mode, the focus of *agility* builds upon the *flexibility* foundation laid by the implementation of ERP, advance planning, supplier relationship management, e-hubs, and a host of other systems. Flexibility, in its most basic sense, means exchanging information, knowledge, intelligence, and optimization results with trading partners about all aspects of supply and demand fluctuations and responding to those changes. However, there is the problem of lag in reaction time and that is where agility and a more adaptive network are needed. The difference between agility and flexibility is that agility includes every component of flexibility, with the addition of speed. Businesses today need a new way of operating that gives them the ability to respond quickly to unexpected changes. An adaptive manufacturing system is designed to help businesses quickly respond to changing market conditions by capitalizing on the strengths of operating units within the company as well as trading, partners outside the company. The optimal KM/WM system model is based on mutual goal setting and not limited to the traditional buyer-seller relationship that exists today. The bottom line is that all parties are engaged in a win-win situation (Baljko, 2003).

Related to the lean inventory techniques is the concept of *continuous improvement,* which is an important element of total quality management (TQM). Continuous improvement is what its name implies — making small, incremental improvements to products over time to meet customers' changing needs. Similarly, customer service improvements provide a competitive edge in industries where fast response times are critical. For example, hospitals require shipments of replacement parts because they cannot afford to have their MRI machines down for long.

An important manufacturing principle relates to a company's *benchmarking* of its own performance in logistics and other areas. For example, a company can track on-time delivery rates, that is, how fast it delivers its products. It can assess its accuracy rate. Delivery of the wrong product on time pleases no one. It can measure cycle time by recording the number of inventory turns achieved in a year with the view of cutting inventories in all of the right places. Logistics costs as a percentage of sales can also be tracked by the company. In all of these benchmarking examples, a company can compare itself to its competitors and companies in other fields in order to get a handle on how the company is performing.

A discussion of newer manufacturing principles would not be complete without reference to a venture analysis modeling approach. As stressed in prior chapters, this approach examines the revenue, manufacturing and related costs, and profit factors over the lifecycle of a typical product or service. A comparison of one product or service to others indicates which product or service should be initiated so that an organization can optimize its resources over time. As with any new technology, there are major organizational and cultural issues involved in getting decision makers to embrace new tools and, more significantly, new ways of conducting business. The bottom line is that a venture analysis modeling approach has the potential to make a marked impact not only on manufacturing, but also on all of the other major functional areas of an organization. All of the preceding newer management principles relating to manufacturing are set forth in Figure 1.

Figure 1. Newer management principles help manufacturing decision makers get an enlarged view for connecting "points of wisdom" about what needs to be done over time

- The whole development and design process centers on including customers, suppliers, manufacturers, wholesalers, and retailers in the total supply chain so that wise decisions can be made by all parties.
- The need for trust among members in the total supply chain is a must. The lack of trust could result in the company's inability to leverage its wide range of technologies not only for itself, but also for other members in the supply chain.
- Excellence in manufacturing means achieving it in quality, reliability, and efficiency. Manufacturing excellence needs to be undertaken on a day-to-day basis.
- The focus of agility builds upon the flexibility foundation laid by the implementation of ERP, advanced planning, supplier relationship management, E-hubs, and a host of other systems. Flexibility means exchanging information, knowledge, and intelligence with trading partners so that all parties can adapt to changing conditions.
- The concept of continuous improvement which is an integral part of the TQM approach centers on making small, incremental improvements to a company's products and services over time to meet the changing needs of customers. Benchmarking is an effective means of comparing one company against its competitors and companies in other fields in order to better understand how the company is performing - good, bad, or indifferent.
- Benchmarking is an effective means of comparing one company against its competitors and companies in other fields in order to better understand how the company is performing.
- A venture analysis modeling approach is an effective way to relate manufacturing operations and cost factors to pricing factors over the lifecycle of a product or service.

Development of an Effective Manufacturing Model for Optimal KM/WM Systems

Development of an effective manufacturing model for optimal KM/WM systems recognizes the importance of a *proactive* viewpoint. More specifically, proactive manufacturers are going to their customers and offering a value-added services approach, such as managing online inventories, as opposed to waiting for customers to ask. This upward collaboration approach requires e-commerce connectors and tools. For instance, Corning, a leading manufacturer of fiber-optics components, is doing this with its customers, Lucent and Nortel.

Such companies need e-business software solutions, such as the eSupplier Collaboration and Flow Production, to support collaborative product design and replenishment to actual demand rather than operating at arm's length. Demand models, volume, and mix are changing so fast that manufacturers need to collaborate on design in order to respond rapidly and stay ahead of requirements (Navas, 2000). As noted previously in the text, an effective manufacturing model for an optimal KM/WM system needs to be taken into account a venture analysis modeling approach. Essentially, this approach integrates customers, product design, marketing, manufacturing, and suppliers from the outset to a point when the product is taken off the market. In view of the above comments concerning an effective manufacturing model within an optimal KM/WM system environment, a number of areas are explored below in some depth. They include: (1) encourage an e-learning organizational environment, (2) utilize lean manufacturing principles to improve operations, (3) optimize e-procurement and supply chain management, (4) utilize total quality management (TQM) in the manufacturing process, and (5) employ manufacturing software useful in optimal KM/WM systems.

Encourage an E-Learning Organizational Environment

An effective manufacturing operation should have as its focus the concept of being a learning organization. Today, progressive manufacturing firms employ e-learning, also known as online learning, computer-based training, distance learning, and on-demand learning. E-learning comes to users via the Internet as opposed to the past where the focus was on the control organization method in a manufacturing environment. While the *control organization method* centered on structured manufacturing operations where most contingencies were anticipated, the *e-learning organization method* is suitable when problem recognition, definition, and solution are likely to differ for most situations. Problem finding can be an effective means of defining future opportunities and problems within an e-learning environment. Hence, there is need for the manufacturing organization to learn over time to adapt to these changing situations. An e-learning organization method centers on organization personnel expanding their capacity to create the result they truly desire over time, utilizing new and expansive patterns of thinking that are encouraged, and continually learning how to learn and work together better. This method centers basically on knowing what the organization and organizational members do well, learning from that in order to do it better the next time, and continually looking for improvements using some type of e-learning approach.

There are a number of e-learning vendors today, which include Caliber Learning, Centra Software, Click2Learn.com, DigitalThink, IBM E-Learning Services, and SmartForce. E-learning offers significant cost reductions and not just in travel expenses. For example, in information, system projects, the largest impact of learning occurs from the person's time away from the job. With Web-based training, individuals can fit training into their work schedules by taking courses at their desks instead of undertaking off-site training. Typically, e-learning is growing largely on the strength of advances in networking and Web technologies, collaboration software, multimedia, content management systems, and other building blocks for educational and training applications. E-learning tools and practices are increasingly coming into play in the training of personnel not only within the organization, but also outside it. In summary, e-learning is not a panacea, but rather a tool or set of tools that can augment other forms of training. Equally important is the fact that e-learning is needed to keep a company and its trading partners abreast of changes that are necessitated by changed business models in order for them to grow their optimized organizations.

Utilize Lean Manufacturing Principles to Improve Operations

Related to an efficient and effective manufacturing environment is the application of lean principles. These principles, which have been around for several decades, focus on getting rid of any and all waste. Lean may have gotten its start on the shop floor, but its real value comes from something much larger: deep integration between shop floor processes, the internal supply chain, and the larger external supply chain environment in which every manufacturer operates.

The promise of a truly integrated lean enterprise is now driving companies of all sizes to reassess how they look at traditional ERP and MES systems. While ERP and MES systems have been instrumental in creating efficiency and driving cost-effective processes at manufacturers in a variety of industries, new thinking along with new technology are needed to make the shift to lean. Lean computer technology environment by definition must be implemented to fit the precise needs of the individual manufacturer. From this perspective, lean manufacturing along with a lean supply chain approach is a belief that everything can be improved — things that are good can get better over time. The bottom line is that manufacturers should look upon lean as a continuous process for improvement.

Optimize E-Procurement and Supply Chain Management

Understanding the e-procurement and supply chain approach can be accomplished by having manufacturing managers work with their purchasing counterparts to evaluate critical purchasing areas using smart software, optimization techniques, multidimensional analysis, knowledge discovery, and business intelligence. In addition, assessments of outsourcing among multiple vendors, analysis of purchasing under uncertainty, analysis of availability of raw materials and parts to meet present and future growth, and the capability of obtaining raw materials and parts for new products being developed can be undertaken. In terms of day-to-day operations, critical purchasing areas that also can be evaluated include the measurement of idle machines and/or personnel resulting from a lack of purchased supplies, the measurement of successful substitutes of materials and parts, the ratio of rejected purchases to total purchases, and the savings on discounts and quantity purchases. By paying attention to the foregoing areas, purchasing and manufacturing managers are provided the means to have better control over their operations as times change.

Today, e-procurement is moving from easy-to-automate tasks like invoice generation to more complex concerns such as e-marketplace implementation, operation, and electronic collaboration. Although the earliest trend in e-procurement adoption was the use of tools to automate processes related to the purchasing of *indirect goods*, the focus today is on technologies that support e-procurement of *direct goods* that are mission critical, such as manufacturing components or raw materials that go into production of finished goods. Because the dollar value of direct materials may account for 80% of some manufacturers' expenses, the savings derived from implementing e-procurement can be compelling. Among the challenges is managing e-catalogs, providing employees with the right mix of products, and making it easy to search for items. Typically, e-procurement technology offers lower purchase prices and faster fulfillment cycles as well as lower administrative overhead. As a result, companies realize broad, measurable benefits from the implementation of e-procurement.

A supply chain approach like e-procurement extends a manufacturer's capability to work with its customers and suppliers on a daily basis. Companies that use the Internet and other computer technology to swap information and knowledge as well as new ideas routinely with suppliers create an environment for joint achievement. By sharing such items and their resulting analyses about customer demand, product defect rates, and engineering changes, manufacturing cycle times and inventory levels can be reduced as well as better products can be built. Leading-edge supply chains can build mutual trust and loyalty.

Companies, like DaimlerChrysler, are currently concentrating less on purchasing and more on managing the supply chain and managing the flow of material from the chain into its plants. Large current savings have come from this approach and the same is expected in the future. At DaimlerChrysler, those who will supply components for its upcoming minivan line are already on line. DaimlerChrysler can tell its suppliers major changes it wants to make in the supply base and set up the sourcing and presource it without having even conceptualized the vehicle. Thus, companies need to have an effective purchasing supply-chain strategy in place for supplying their future manufacturing activities.

The tie-in of purchasing activities with a supply-chain management approach via the Internet is the realization that the purchasing of goods and services on the Internet is faster, more efficient, and more cost effective for manufacturers as well as non-manufacturers. The Internet provides organizations with complete line-item detail on each transaction, giving users real-time access to critical product information, knowledge, intelligence, and optimization results. Additionally, the purchasing and supply-chain structure is linked with finance to ensure that there is an easy flow between these organizational functions.

Data-mining tools (as set forth in Chapter IV) can be utilized to assist in the development of an optimum e-procurement and supply chain strategy by discovering and possibly discerning upcoming trends, patterns, and relationships that lie hidden in data warehouses (extracted from intranets and extranets). These facts, in turn, can be disseminated via the Internet. In addition to utilizing information, knowledge, intelligence, and optimization results from these sources to ensure that the best opportunities in terms of products and services are achieved, the discovery of new facts and a better understanding of manufacturing operations can ensure that an optimum e-procurement and supply chain strategy is being pursued. Typically, a most important part of this strategy is about developing the right kind of procurement process, understanding this procurement process, and making sure it pays for itself.

Utilize Total Quality Management (TQM) in the Manufacturing Process

Today, quality has become a service issue — not just for service sector businesses like communications, health care, and finance, but for the service side of manufacturing firms as well. The focus is on *total quality management* (TQM) at a higher level where quality in the offering itself and in all the services that come with it, starting with the design process. When the growing popularity of Japanese automobiles, televisions, and radios forced U.S. manufacturers to take another look at themselves, in the late 70s, most companies were still in what quality experts call the first, that is, inspection, phase. That is, the focus was on sampling techniques to get rid of defective items. Too often, however, they did not. In 1980, an NBC White Paper introduced audiences to W. Edwards Deming, the American statistician who had shown the Japanese how to use process controls to catch defects at the source. Manufacturers who took the issue seriously started moving into the second or quality control phase. Now, with TQM or the third phase, quality is no longer solely in the quality-control department since it is sponsored by top management and diffused throughout the company.

Today an important part of total quality management is a tie-in with optimal KM/WM systems to grow the leaning organization. For example, consider the automobile industry, where quality is still defined as the absence of manufacturing defects. In the years gone by when there was a large gap between the product quality of American and Japanese vehicles, that was certainly a valid measurement because it had a direct impact on customer satisfaction. Today, the difference in defect rates among the major automakers is somewhat indiscernible to customers. Typically, research indicates that customers feel that a good measure of vehicle quality is a balance between minimizing the *things gone wrong* with those attributes that customers consider *things gone right*. One way of bringing these two areas together is the utilization of mathematical and statistical analytics within an optimal KM/WM operating mode for a company's TQM program. Additionally, this customer output can be used to get some insight into examining in more detail TQM from a worker's viewpoint that starts with the design process. Perhaps, both the customer's and worker's viewpoints can be merged together such that there can be a win-win situation for better optimizing a company's operations from within and outside the company. In summary, TQM has become a prerequisite for survival today and tomorrow for creating and distributing unique values of quality products and services to a company's customers that is self evident from the very beginning of the design and manufacturing processes.

Employ Manufacturing Software Useful in Optimal KM/WM Systems

Today, there are a number of software packages helpful in the development of manufacturing systems within an optimal KM/WM system environment. They include the following: (1) optimization, (2) supply chain management, (3) enterprise resource planning, (4) manufacturing execution systems, and (5) artificial intelligence. All of these topical software areas are treated below. In addition to the software packages, reference can be made to Chapter VI (corporate planning) and Chapter VII (marketing) for product lifecycle management software. Arena Solutions, for example, is offering a Web-based, hosted approach to PLM, which is typically better than a client/server approach. In a client/server environment, local information knowledge and intelligence is trapped behind a firewall. In contrast, with PLM software as a service, there is on controlled central environment on the Internet where supply chain operations can operate on an optimum basis (Stackpole, 2005). As noted previously in the text, PLM software is a way to manage a product from the dawn to dusk of its life: from creation through development, manufacturing, testing, and then maintenance in the field. Expansion of PLM over the life of a product (or service) results is an expanded approach set forth previously in the text as venture analysis modeling.

Whether a company's operations are viewed on a local or a global basis, it is helpful to utilize some type of *optimization software*. As a starting point, improved forecasting demand on a local and global basis both in the inbound and outbound supply chains is a key component. There is need to deliver goods faster and more efficiently as well as to wield the supply chain as a tool of corporate strategy. This comprehensive approach to a company's operations requires a very broad approach to optimizing its forecasting, manufacturing, and supply chain constraints. In some cases, statistical models, such as time series and regression analysis, may be appropriate ones while, in other cases, they may not be to get a real handle on the company's forecasts. Also, mathematical models, such as linear programming and its derivatives (i.e., integer linear programming and non-linear programming), can be useful for meeting a company's manufacturing constraints. In terms of the whole supply-chain approach, optimization technology can be useful to determine optimum or near-optimum monthly production schedules, which can answer such questions as to which factories should make what products to serve local and global customer needs.

For improved manufacturing operations, Nissan turned to constraint-based programming and optimization software. This software takes into account all the manufacturing constraints while making the best use of production resources to arrive at a solution. At Nissan's Sunderland plant, each of the two production lines has a body, paint/trim, and chassis shop. On each line, the first body is made while the second is painted and the third car has the parts fitted. There are literally billions of combinations in which vehicles can be placed, each of which might yield a feasible solution. Nissan chose a British consultancy company, PA Consulting, to do the constraint-based programming using optimization software from Mountain View, California-based iLog, including iLog Views and iLog Solver. The iLog's tools linked all of the zones together and scheduled them in a timed sequence. Essentially, it is a rules-based system that knows how to sequence the vehicles. Then iLog Solver, the engine, is used to come up with an optimum solution (Schwartz, 2001).

Today, companies need *supply chain management software* that provides the right information and knowledge at exactly the right time and place for efficient and effective manufacturing operations. The leading software vendors in this area are i2 Technologies, Manugistics Group, Oracle, PeopleSoft, and SAP *AG*. These software vendors are improving their products to perform in real time, across heterogeneous environments, and with increased analytics capabilities. Their real-time capabilities allow for better integration with other applications and map more closely to the business processes and priorities on which companies actually operate. The responsiveness that real-time visibility offers allows companies the capability to operate with lower inventory levels. Real-time visibility into their supply chain improves forecasting capabilities, which is important because inaccurate forecasts translate into lost sales.

In addition, supply chain management software vendors have added support for Web services so their applications can more easily share information and knowledge with other vendors' applications as well as with legacy systems. Some vendors have rebuilt their software so it is purely Web-based. Such efforts have set the stage for companies to automate business processes across departments and beyond their own operations (Bacheldor, 2003). The applications themselves continue to evolve so companies can make better use of the data that supply chains generate. New analytical functions built into supply chain software can help users understand how well a supplier performed without having to open up another application, for instance. There are also new features that automatically send alerts — via e-mail, cell phone, and other devices — when things are not going well.

As in the past, the primary goal today of *enterprise resource planning* (ERP) is related to improving customer service: getting the right product in the right quantity to the right place at the right time in the right condition and at the right price. Essentially, ERP is real-time manufacturing software that incorporates computer supply chain management software from sales and order to production, inventory, and distribution. In as much as companies are implementing systems that manage the entire supply chain (from the time a product order comes in to the moment it is delivered), ERP systems represent an integrated set of applications modules that ties together a host of functions. They are highly integrated application suites that support many processes, including sales forecasting, order management, purchasing, production scheduling, inventory management, distribution, scheduling, engineering, maintenance, and accounting. These systems, then, are the nerve center of a manufacturer. Enterprise resource planning software is available from a wide range of vendors who have the capability to provide competitive advantage through lower costs and faster response time. ERP vendors are Baan, J.D. Edwards, Oracle, PeopleSoft, and SAP *A*G.

While manufacturing software, such as ERP, represents a broad-based approach to manufacturing, *manufacturing execution systems* (MESs) applications guide manufacturing processes through each step and collect information about what was done, to what, by whom, and on which machine. Detailed instructions and shop-floor data capture allow other applications to analyze what has already happened and what needs to happen next. As more and more manufacturers have come to realize about MESs, it is not enough to have information. They need to maximize the value of that information by incorporating knowledge management and business intelligence into their product suites. Tecnomatix, for example, has moved its client/server based MES application over to a Web-based version. Datasweep Inc. is marketing an Internet-based

MES product, called Advantage, that uses predictive analysis. This newer approach positions MES as a business facilitator because it can gather information from anywhere in the world, collect it in a database to be analyzed, and offer interpretations that enhance enterprise performance. Other MES players like Camstar Inc. have added analytical tools to its InSite product suite, and control vendor Rockwell Automation has branched out with MES software that can do overall equipment effectiveness (OEE) calculations.

An important software development applicable to manufacturing is *artificial intelligence* (AI). The advantage of the new ISO 18629 language in AI is that it can be revised and employed to link a variety of applications. This new software language was developed by federal government researchers at the National Institute of Standards and Technology and colleagues in France, Germany, Japan, and the United Kingdom. This process-specification-language software should make computers reason much more precisely than they do now. In manufacturing, the language is well suited for the exchange of process planning, production planning, and control information and knowledge for guiding manufacturing processes. As such, the ISO 18629 language can be used to build a variety of advanced manufacturing systems as found in an optimal KM/WM system-operating mode. Since the notion of process underlies the entire manufacturing cycle, coordinating the workflow within engineering and shop-floor manufacturing can be difficult. The ISO 18629 PSL (product system language) software employs artificial intelligence and language analysis to denote computer commands in the framework of a production plan. Simply put, PSL acts as an interpreter between two applications say, a system that runs a manufacturing process and one that schedules a process, thereby eliminating any ambiguity of the definition of terms used by each system (Chabrow, 2005).

Manufacturing Model and Its Sub-Models Found in Optimal KM/WM Systems

Typically, there is a tie in of the marketing model (as set forth in Chapter VII) to the *manufacturing model* that, in turn, is related to the finance (and accounting) model (in Chapter VIII). The manufacturing model, as set forth in Figure 2, has a number of sub-models and related parts. Essentially, the marketing model and its sub-models set the stage in that a company's products and services add unique values for customers. In turn, these unique values can be integrated into a company's manufacturing model and its sub-models such that there is a win-win situation for all parties involved. An underlying force

Figure 2. Relationships of the manufacturing model and its sub-models to the other functional models (sub-models) within an optimal KM/WM system environment

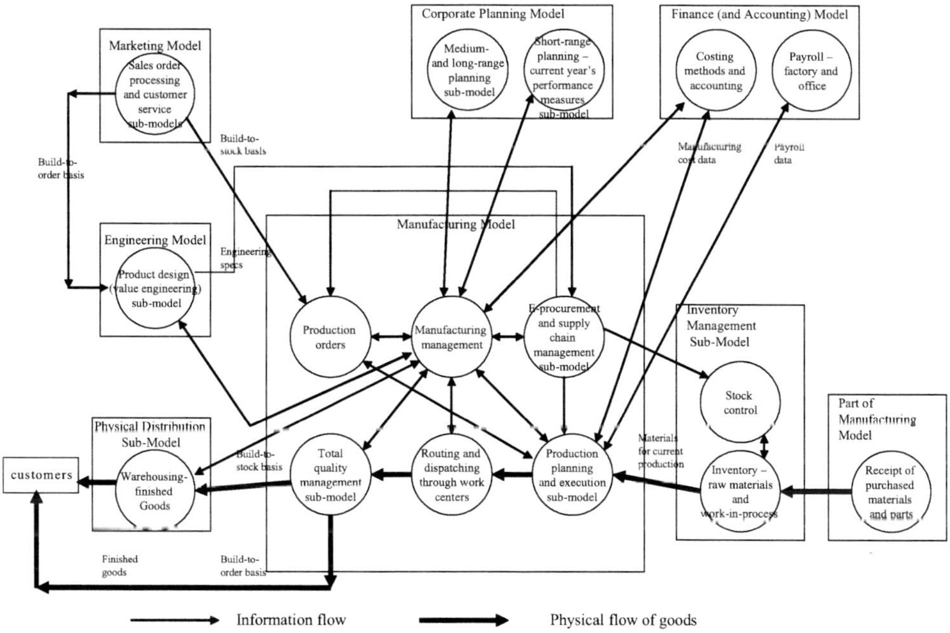

behind this win-win situation is the finance (and accounting) model that assists in the optimization of a company's resources.

As a starting point, the *production planning and execution sub-model* — as shown in Figure 2 — determines what products should be made on the production floor for a build-to-stock basis and a built-to-order basis. The specifics of both bases will be given later in the chapter. It is sufficient to say at this point that enterprise resource planning (ERP) is employed as a broad-based approach to the manufacturing process that, in turn, is guided by a manufacturing execution system (MES) that guides manufacturing activities on the production floor. As shown in the illustration, raw materials and work-in-process provide the materials for current production.

To assist this first sub-model, manufacturing management needs to employ the *e-procurement and supply chain management sub-model*. Once raw materials and parts have been purchased on an optimum basis, management needs to evaluate vendor and buyer performance as well as purchased performance indexes (PPIs) on a periodic basis. This evaluation process allows manufacturing decision makers to make changes to the company's supply chain over time. In addition, management needs to use value analysis or value engineering for improving the company's products over time.

Typically, the need to keep a company's investment in inventory at a low level is paramount. As expected, there is a tie in of the third sub-model, that is, *inventory management sub-model*, with e-procurement and supply chain management. Basically, there are two ways to keep inventories under control. One is the employment of statistical and mathematical models to buy on an optimum basis, so that there is an even flow of inventory into company's warehouses and out into the manufacturing process. The other way is to employ the just-in-time (JIT) concept to minimize a company's investment in inventory. Going one step further, there is also need for a manufacturer to examine the benefits of cooperating on an inventory management program with its customers, trading partners, and employees.

The last of the manufacturing sub-models is the *total quality management* (TQM) *sub-model*. As indicated in Figure 2, TQM is an integral part of the design and manufacturing processes for a typical company. At the various stages of the design and manufacturing processes, appropriate measures, that is, statistical and mathematical analyses, can be used to gauge acceptable versus non-acceptable performance, that is, defects in the actual manufacturing process. When deemed necessary, corrective action is undertaken to eliminate known manufacturing defects. In addition, it is helpful to tie in TQM throughout the organization in order to maintain the optimized organization. All in all, the activities of these four manufacturing sub-models are interrelated and they provide a basis for evaluation of manufacturing operations, their costs, and important manufacturing measures and ratios.

Production Planning and Execution

Because optimization can be an effective way to assist in making sound manufacturing decisions on a day-to-day basis over time, it behooves manufacturing decision makers to utilize all types of technology in production planning and execution on the factory floor. This includes utilizing a wide-ranging manufacturing database, which consists of various elements, such as upcoming forecasts along with manufacturing strategies, facts, heuristics (rules of thumb), and procedures. Additionally, modeling capabilities using mathematical and statistical techniques can assist manufacturing managers and their staffs in optimizing its manufacturing operations. Of equal importance is available information and knowledge from a number of sub-models (refer to Figure 2), product lifecycle management, enterprise resource planning, e-procurement, supply chain management, and inventory management which can be combined with current manufacturing information, knowledge, and intelligence to provide a better understanding for optimizing manufacturing operations so that wise decisions can be made. This approach is taken below for production planning and execution on the factory floor. All in all, information, knowledge, intelligence, and optimization results, which are the backbone of manufacturing operations, need to be recognized by manufacturing decision makers as the key to adding values for customers, trading partners, and employees.

Tie-In of Production Planning to Corporate Planning for the "Big Picture"

Production planning and execution, by itself, is not capable of providing the "big picture" for an optimal KM/WM system-operating mode. It must be related to corporate planning so that an organization's vast experiences over time can be defined and fine tuned to meet the challenges of current and future manufacturing needs. As noted previously in the text, a venture analysis modeling approach, with accent on PLM, has the capability of integrating corporate planning, that is, capital projects, with the manufacture of products over their lifecycles. Hence, the merger of corporate planning with production planning and execution is an effective means for offering sound advice to decision makers at all levels so that profits are maximized over the long run.

At a more practical level, the production planning and execution department is responsible for all physical movements between manufacturing departments and within their respective work centers at the plant level. This department coordinates all activities concerning a production order from its initial recording, through inventory control and manufacturing, to getting the finished goods ready for shipment to customers, that is, direct shipments, or to company warehouses for subsequent shipment to customers. Production planning and execution is dependent on the plant's database and its communications with all manufacturing work centers. Daily manufacturing activities, which are the responsibility of the production planning and execution department, need to be analyzed continuously by manufacturing managers and their staffs for efficiency and effectiveness. Also, there is generally need to discover and track manufacturing problems using a TQM approach. From this perspective, manufacturing managers and their staffs are assured that short-range manufacturing plans are being implemented on a current basis as mandated by manufacturing management.

Initially, manufacturing partnerships and their relationship with production planning and execution are set forth below along with connecting "points of wisdom" in the manufacturing process. Additionally, determining the appropriate production planning and execution techniques for current operations on a build-to-stock basis and a build-to-order basis are covered below. It is essential for manufacturing management and their staff to link manufacturing with corporate planning so that appropriate optimization techniques that really optimize a company's manufacturing operations are implemented over time. In this manner, the right hand knows what the left hand is doing from the standpoint of connecting points of wisdom for making wise manufacturing decisions.

Focus on Optimizing Trading Relationships that Tie-In with Production Planning and Execution

Supply chain management has been and is widely practiced today. For example, in the past, many large manufacturers, particularly in Japan, formed *keiretsu* or company partnerships by grouping the subcontractors who supply them with parts. Such a production system, distributed among many firms, has helped manufacturers strengthen their global competitiveness. More recently in Japan, however, the subcontracting system has changed to reflect the structural sophistication in industry. Automakers, electric machinery manufacturers, and other large Japanese firms, which are suffering from sagging demand and decreasing earnings, are increasingly pressuring subcontractors to cut production costs. Non-compliance can result in being excluded from the keiretsu.

Because many subcontractors are looking for business partners outside their company groups, this presents a good opportunity for subcontractors to become independent. In effect, the keiretsu system is evolving from the existing pyramid-type coalition to horizontal or network type coalitions. An increasing number of subcontractors are seeking to improve their technology and production to satisfy key manufacturers. While some are pursuing new business partners outside their company groups, others aim to become independent specialists in processing or manufacturing. For example, because U.S. car manufacturers (as noted previously) are heavily involved in sophisticated supply-chain management systems, they include a focus on various cost-cutting programs, including reduction of work hours and model lines, extension of model-change cycles, and increased use of common parts in the manufacturing process.

Within established manufacturing supply-chain partnerships, the focus now becomes one determining the appropriate planning and execution technique over manufacturing operations. *Enterprise resource planning* (ERP) which is a comprehensive approach to manufacturing utilizes real-time manufacturing software that incorporates the entire supply chain — from sales and order processing to production, inventory, and distribution. To assist in reducing operational errors between forecasts and actual numbers, the best approach is to manage the *error in demand planning*. At the lower level is a *manufacturing execution system* (MES) that utilizes software to manage shop floor activities. An integral part of MES is generally *just-in-time* (JIT) *inventory*. JIT, which is based on the elimination of wasteful or non-value-added activity within a company, aims to reduce inventory as close to zero as possible and to reduce lead-time via reduced setup and reduced lot sizes in the manufacturing process.

Recently, the Instrumentation, Systems and Automotive Society (ISA) has defined a standard S95 that fills the communication gap between the enterprise and manufacturing. Essentially, S95 defines the terminology between an ERP system and a MES framework. Part of that framework relies heavily on the World Batch Forum (WBF) standard, called the business-to-manufacturing markup language (B2MML), which serves up the appropriate context around the data. As long as the ERP system and the MES system both comply with these standards, they speak the same language.

Regardless of the industry, manufacturers need to address quality, traceability for regulatory compliance, and global operations management. An integral part of these issues is *integration*, first in the enterprise and then in the supply chain. Additionally, companies need to tie product lifecycle management (PLM) applications together with manufacturing operations. PLM gives the manufacturers the capability to take a 3D product design simulation from a CAD/CAM package and move it directly to the factory floor as a reference model for streamlining the process between engineering and production.

All of these requirements ushered in a new wave of MES. The ISA S95 committee is using *manufacturing operations management* (MOM) as a new way of viewing manufacturing operations. Most MES vendors have reengineered client/server products around the Web technology using Microsoft's .Net technology or the J2EE development platform. This allows the MES application to scale to accommodate worldwide use as well as to invoke Web services that add functionality. As part of the Internet-enabled redesign effort, most MES applications are also tapping into ISA's S95. Overall, the move to S95 and B2MML is a large step forward for resolving the issues of integration (Neil, 2005).

Connect Manufacturing Operations to "Points of Wisdom" for Better Optimization

An integral part of adding unique values to customers over time is the whole concept of making a company's customers its partners so that "points of wisdom" can be connected for optimization of a company's operations as well as for its customers. This collaboration approach is quite at home within a manufacturing operating mode. In order to accomplish this collaboration in manufacturing, the practice of *rethinking* needs to be applied before launching any product into the manufacturing process. Rethinking can be applied to abstract concepts as well as to the more concrete realities. For instance, conventional wisdom may have influenced the first manufacturing database, but such a database is light-years beyond a manual approach. The relational side of a manufacturing database makes cross-referencing parts and raw materials to the manufacturing process a taken-for-granted item today. The same logic can be applied to manufacturing operations in the future where the focus will be on connecting points of wisdom within manufacturing departments as well as with other functional areas.

To assist in the collaboration and integration effort needed by manufacturing decision makers to connect points of wisdom on a build to-stock basis and a build-to-order basis, reference can be made to Figure 3. As a starting point, a *cause-and-effect diagram* can be drawn that centers on scheduling of daily activities where wisdom can be applied for optimizing a company's manufacturing operations. More specifically, on a build-to-stock basis, questions can be asked, such as which order or orders should be manufactured first. That is, which product or products need to be produced now to replace stock, especially if items are backordered or continually in short supply. Similarly, are the costs of the items produced about the same or completely different that affect the capital invested in inventory? From this perspective, a root or basic product can be determined such that all inventory considerations are taken into account. Essentially, this approach to a build-to-stock basis places first things first for

Figure 3. Employment of an appropriate production system to plan, execute, and evaluate manufacturing operations on a build-to-stock basis and a build-to-order basis

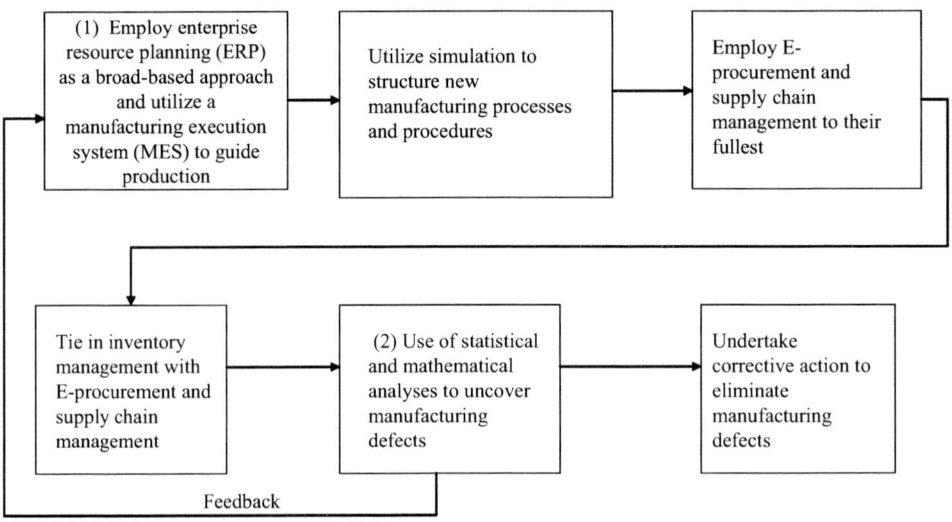

making wise manufacturing decisions on a daily basis. In terms of a build-to-order basis, there is generally little need to develop a cause-and-effect diagram.

Once a cause-and-effect diagram approach is completed, the next step is to connect points of wisdom that are essential to the optimization of manufacturing operations. As shown in Figure 3, the important points of wisdom that need to be connected for effective employment of manufacturing resources are: (1) employ enterprise resource planning (ERP) as a broad-based approach to the manufacturing process and utilize a manufacturing execution system (MES) to guide production and (2) use statistical and mathematical analyses to uncover manufacturing defects. The connect between these two important points of wisdom in Figure 3 permits a company's manufacturing decision makers to get important information, knowledge, intelligence, and optimization results necessary to maintain efficient and effective production operations on a daily basis.

To assist manufacturing decision makers in connecting other points of wisdom, reference can be made to Figure 4. For the first basic manufacturing function — *production planning and execution* — there are four detailed activities, the first and third were set forth above. Basically, these detailed activities ultimately center on getting customers the right product at the right time and place. The second detailed activity — differentiate between a build-to-stock basis and a build-to-order basis — relates to the real essence of manufacturing operations on a daily basis. That is, are products manufactured to meet a specific customer order or are they made for stock and, in turn, for subsequent shipment to customers? The fourth and last detailed activity — utilize simulation to structure new manufacturing processes and procedures — employs a 3-D modeling approach to better understand key manufacturing equipment and their processes for current and planned production operations. Similarly, the other basic manufacturing functions — as noted in Figure 4 — are e-procurement and supply chain management, inventory management, and total quality management (TQM). As before, these basic manufacturing functions can be related to their detailed activities, which can be connected to points of wisdom for better optimizing them. It should be noted that the two major points of wisdom in Figure 3 are related to first and third detailed activities for production planning and execution as well as the second activity for total quality management found in Figure 4.

Make Production Planning and Execution an Integral Part of Product Lifecycle Management

Related to an effective production planning and execution approach to manufacturing products is the product lifecycle management operating mode. This includes engineering the product to meet specific requirements as deemed necessary by the customer and the company's production facilities whether they

Figure 4. Relationship of a company's basic manufacturing functions and detailed activities to connecting "points of wisdom" for their measurement and analysis

Basic Manufacturing Functions (Figure 2)	Detailed Activities of Basic Manufacturing Functions	Connect "Points of Wisdom" For Optimal Decisions in Manufacturing	Measurement and Analysis Needed for Optimal Manufacturing Decisions
Production Planning and Execution	(1) Employ enterprise resource planning (ERP) as a broad-based approach to the manufacturing process	Centers on improving customer service by getting them the right product at the right time and place	ERP analyses
	(2) Differentiate between a build-to-stock basis and a build-to-order basis	Relates to the real essence of daily manufacturing operations	Build-to-stock and build-to-order analyses
	(3) Utilize a manufacturing execution system (MES) to guide production	Collect information about what was done by whom and on what machines	MES analyses
	(4) Utilize simulation to structure new manufacturing processes and procedures	Employ a 3D modeling approach to key manufacturing operations	Simulation analyses
E-Procurement and Supply Chain Management	(1) Evaluate vendor and buyer performance periodically	Relate vendor to buyer and vice versa for recognizing excellence	Vendor and buyer analyses
	(2) Evaluate purchased parts performance as to price, quality, and delivery	Relate those factors that matter for optimizing the supply chain	PPI analyses
	(3) Make appropriate changes to the company's supply chain over time	Make changes to keep the supply chain effective on a daily basis	Cost/benefit analysis
	(4) Use value analysis or value engineering for product improvements	Recognize that product improvements are needed over time	Content analysis

Figure 4. continued

Basic Manufacturing Functions (Figure 9.2)		Detailed Activities of Basic Manufacturing Functions	Connect "Points of Wisdom" For Optimal Decisions in Manufacturing	Measurement and Analysis Needed for Optimal Manufacturing Decisions
Inventory Management	(1)	Tie in of inventory management with E-procurement and supply chain management for compatibility purposes	Collaboration of supply chain factors with inventory factors	Cost/benefit analysis
	(2)	Employ quantitative methods to buy on an optimum basis for manufactured and stocked items	Utilize the computer for automatic reordering	EOQ (economic ordering quantity)
	(3)	Utilize the just-in-time (JIT) concept to minimize investment in inventory	Employ optimization techniques to determine JIT	Inventory levels analysis
	(4)	Examine the benefits of cooperating with others for inventory management	Collaborative efforts to reduce inventory warehousing costs	Cost/benefit analysis
Total Quality Management (TQM)	(1)	Integration of TQM throughout the design and manufacturing processes	Make TQM an integral part of all aspects of a new product/service	Quality evaluation analyses
	(2)	Use of statistical and mathematical analyses to uncover manufacturing defects	Employ appropriate analyses before, during, and after manufacturing	Product defect analysis
	(3)	Undertake corrective action to eliminate manufacturing defects	Examine known manufacturing defects by knowledgeable personnel	Zero defects analysis
	(4)	Tie in of TQM with the organization's business model to monitor the optimized organization	Enlarge TQM by going outside its traditional boundaries	Organization business model analysis

Figure 9.4 Continued

are small, mid-size, or large size. As noted previously in the text, it is recommended that a Web-based, hosted approach be used for PLM and venture analysis modeling approaches. When the engineering process is completed, manufacturing management needs to apply the latest optimization techniques on a build-to-stock basis or a build-to-order basis. These two bases are treated in the next section of the chapter.

It is sufficient to say that these techniques are employed such that the essential elements of manufacturing are integrated with corporate planning, marketing, and finance. For example, some companies are implementing an enlarged version of ERP, which includes profit management automation (PMA). PMA combines profit planning, activity-based costing, treasury, and audit management into an integrated suite. PMA gives a company the same sort of control over capital that traditional ERP provides regarding transactional data. In addition, PMA can be tied-in with the PLM approach on the factory floor (Leibs, 2003).

Essentially, the outputs from the build-to-stock basis or the build-to-order basis give company decision makers the ability to analyze manufacturing operations as an integral part of the PLM and PMA approaches. More specifically, a number of performance measures can be developed: (1) customer relationship and service; (2) supply chain; (3) manufacturing operations; (4) data mining, OLAP (online analytical processing), and query reporting; and (5) scorecard financial analytics. The net result of these comprehensive analytics is to either continue or discontinue production of current products and related services. Overall, the essential elements of production planning and execution provide an underlying ERP framework that includes PLM and PMA.

Apply Latest Optimization Techniques for a Build-to-Stock Basis and a Build-to-Order Basis

A company's products can be produced upon receipt of customers' orders, in anticipation of demand, or in some combination of the two. If goods are being produced to order, that is, "build-to-order" basis, the usual arrangement is to have the production planning and execution department initiate action on the order via its data communications network, which can be distributed on line to stock control, shipping, and accounting departments. If goods are produced in anticipation of demand, they are produced on a "build-to-stock" basis. This is the approach taken below where the focus is on planning manufacturing activities on a monthly basis at the corporate level, then controlling on a daily basis using one of the current software packages at the manufacturing plant(s).

On a *build-to-stock* basis, the procedure is to determine next-period's sales forecast (one month hence) by first adjusting for finished goods on order and on hand — shown in Figure 5 — for online production planning. For manufacturers to be able to meet customers' ASAP (as soon as possible) requirements, mapping the error in demand planning is necessary. Top-down demand forecasting systems, for the most part, exacerbate forecast errors since they generally rely on aggregate data, which cannot account for variable demand patterns for individual items. Typically, the forecasting rules built into ERP and manufacturing systems are not adequate; a more rigorous process needs to be incorporated for all lead-time variables, including demand planning and supply variability, to supply optimized inventory for each item at each location. Such a system is the GAINSystems' GAINS*OPS, an ICO (inventory chain optimization) package which takes an expert system, statistical-analysis approach to balancing inventory to achieve optimal cost savings and service levels.

To better stand the ICO system, data from existing ERP systems as well as existing or similar product history are used to create a demand plan for each item that uses multiple forecast models. A series of algorithms analyzes this data to determine an optimal inventory level and cost to meet the most profitable service level. Essentially, the system is regenerative, using a cognitive function to measure its own prediction rate and to self-correct, as well as to adapt to changes in cost, environment, and demand. ICO is able to balance stock by changing the profile of inventories in buffers, thereby reducing overall inventory levels and costs while maintaining or improving service.

Finished-goods production requirements, as shown in Figure 5, for the next month provide the input for manufacturing. Linear programming is used to determine what quantity of each product will be produced in each of the manufacturing plants. Once next-period production schedules by plants have been determined, the next step is "exploding" bills of materials, that is, the materials-by periods program multiples the quantity needed of each component times the number of final products that must be manufactured for the coming month. The component requirements are placed in the appropriate planning period, because some parts will be needed before others. Hence, purchased parts and raw materials are placed on a just-in-time basis for production needs for the coming months.

As shown in Figure 5, the output for the materials requirements by future-planning periods can take one of two paths. The first one is the e-procurement of raw materials and parts from outside vendors, and the second one is the manufacturing of parts within the plants. Essentially, the outside raw materials provide the basic inputs for manufacturing specific parts used in the assembly of the finished product. On the other hand, outside purchased parts are used in the assembly of the final product. Before materials are to be purchased or manufactured on a build to-stock basis, present inventories and materials on order that are capable of meeting the company's needs for future planning periods are determined. At this point, it is important to note that perpetual inventories stored on line have been adjusted to reflect physical counts in order to produce accurate output for the materials-availability and EOQ (economic ordering quantity) program.

Once monthly production quotas have been established, they can be translated into planned daily requirements, that is the MES software provides the means for scheduling, that is, executing, production orders on line through the manufacturing work centers on a daily basis. Other operational programs that are used to record and execute daily activities include attendance, payroll, work-in-process, and so forth. The output of these programs provides operations evaluation analytics on manufacturing activities on a daily basis that, in turn, can be extended over a longer period of time to report information, knowledge, intelligence, and optimization results about what is actually going on in the company's manufacturing operations.

Going one step further, a daily-computerized scheduler is utilized at each plant in order to smooth production for each working day. Before the start of each working day (the program is actually run at the end of the prior day shift and reviewed by the plant's management), the scheduler considers where jobs are backed up or behind schedule and where production bottlenecks are currently occurring. Utilizing these inputs, the scheduler simulates the activities of the plant for the coming day and determines what will happen as the day begins, thereby alerting the plant superintendent and foremen to critical areas that need immediate attention in manufacturing operations. Since manufacturing activities are in real time, the scheduler feeds back information in sufficient time to monitor upcoming manufacturing operations. Where deemed necessary, the daily computerized scheduler allows the production planning and execution department to make adjustments to accommodate last-minute changes.

Going beyond the build-to-stock basis, progressive companies are moving ahead or have moved to the *build-to-order basis* for manufacturing operations. As shown in Figure 5, customer orders are received via the Internet, fax, e-mail, others. Typically, there is need for customization of the order, which may require the services of the engineering department for detailed design and engineering of the product. These specifications provide the input for production planning and execution which results in production scheduling for a build-to-order basis. As also shown in Figure 5, generally, there is a need to order raw materials and parts that are tied-in with the materials planning-by-periods program. In turn, materials requirements for future planning periods become input for the materials availability and EOQ program. The output, like with the build-to-stock basis, is planned requirements for purchased and manufactured parts.

Many companies today are using the build-to-order basis along with the Internet for manufacturing operations. As noted in Chapter I, Dell Computer has become the textbook example of how to run a business on a build-to-order basis. Currently, Toyota, which is one of the largest carmakers in the world, is

Figure 5. A daily build-to-order basis and a monthly build-to-stock basis for a typical manufacturer using the latest optimization techniques

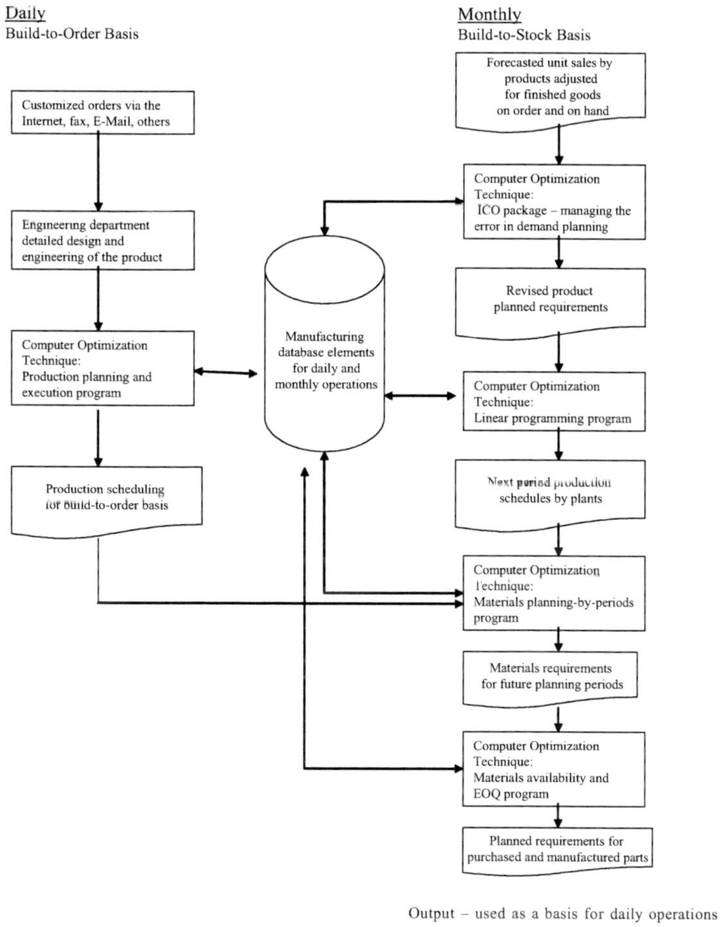

experimenting with a build-to-order basis for its manufacturing operations. The new system would give Toyota an edge in realizing its ability to build an automobile to a customer's specifications and deliver it within days. Such a model would appeal to buyers of all ages. Similarly, DaimlerChrysler is working toward the same goal. The next step for DaimlerChrysler is to add the software that Toyota saw as its starting point — software for daily production and factory-floor modeling, that is, Dassault's Delmia production-support tools to its mix of design-collaboration and product-life-cycle management applications. Different divisions of DaimlerChrysler are at various stages of deploying the software. Use is most widespread at Mercedes and DaimlerChrysler's commercial-vehicle unit, where it is used to model the production process; the Chrysler Group is deploying the software. In total, manufacturers and their customers stand to benefit from using newer software on a build-to-order basis (Konicki, 2002).

It should be noted that Toyota currently has the capability of building a car to a customer's specifications in just 14 days. More specifically, Toyota can tap into the Dealer Pipeline Management System to assemble and deliver special-order vehicles. The system gives a Toyota salesperson a window into the future, that is, what cars will be built for his/her dealership during the coming weeks at the company's four North American plants. If the salesperson wants a vehicle that is not earmarked for his/her showroom, the individual enters in the request and the software will immediately attempt a trade. If such an exchange is not possible, there is still a 90% chance that the desired car will be available within two weeks (Rosmarin, 2004).

Other manufacturers have joined the world of build-to-order for the masses. They include Lands' End, Nike, and M&M's. For example, Lands' End hired Archtype Solutions. Archetype has developed a series of algorithms that pinpoint a person's body size by taking just a few of his/her measurements and then running them against a huge database of typical sizes to create a unique pattern. The customer provides the information on Lands' End's Web site by answering a series of 15 questions (for pants) or 25 questions (for men's shirts) about everything from waist size to inseam. The process can easily take 20 minutes, but since Lands' End saves the data on its Web site, thereby making reordering straightforward. The order zaps to Archtype's office in San Francisco, where its software generates a pattern and sends that to one of five contracted manufacturers in the U.S. and overseas. Those plants cut and sew the material and ship the finished garment directly to the customer. Today, customization is available for most categories of Lands' End clothing. Overall, faster and more ubiquitous Internet technology has made ordering such goods easier. In addition, flexible just-in-time manufacturing methods, coupled with the rapid proliferation of digital technology, have enabled companies to fill individual orders much more quickly than in the past (Schlosser, 2004).

To effect more wise manufacturing decisions, the build-to-stock basis and the build-to-order basis center on the capability to synchronize inventory with demand. Nothing so dramatically affects the bottom line as minimizing inventories of all types. Companies have aggressively pursued strategies such as contract manufacture of all or most of the product, vendor-managed inventory (VMI), and just-in-time (JIT) and Kanban replenishment to meet that goal. As manufacturers adopt build-to-order business models that give customers the ability to tailor orders to their specifications, manufacturing managers are going to have to contemplate changes to the factory floor. One such shift might be changing line-oriented production facilities to work cells, which allow manufacturers to respond more dynamically to product options. Companies may opt to build a line just for e-commerce or create two lines — one for build-to-stock and another for build-to-order.

Related to inventory for a comprehensive management approach is a continuous improvement in manufacturing operations. This centers on a consistent set of procedures and techniques to address price, delivery, service, quality, and time to market. Its elements include determining the cause of material backups, which are often due to combining batches to optimize setups or to combining like-size materials to optimize yield. A helpful approach is to diagram and analyze the movement of parts needed to make an assembly and fulfill shipments to ensure parts will end up in the right place at the right time. To improve work-in-process in a company's manufacturing operations, only material in process for actual orders is worked on while other material is moved to the side. Orders are processed on a first-in, first-out basis by customer due date. To improve the work-in-process flow, machine setup and workflow operations can be videotaped. Tapes can then be analyzed by breaking them down into time elements to spot areas for improvement. It may be necessary to reorganize work areas to improve manufacturing cycle time. From a broad-based and wise view, manufacturing operations may show significant improvements in lead-time reductions, inventory reductions, partial elimination of overtime, reductions in scrap and rework, and the need for training of workers.

To assist in expediting inventory on the factory floor, there are newer smart technologies from software vendors. For example, Pinpoint and WhereNet have developed a RFID (radio-frequency identification) tracking system, dubbed LPS (local positioning system) or RTLS (real-time location system), that uses triangulation — much like the satellite-based GPS (global positioning system) — to locate assets or personnel within a non-line-of-sight range of about 250 feet. For example, Pinpoint's LPS 3D-ID system uses a network of antennas and cell controllers planted within a facility to triangulate on long-range active RFID tags with a battery life of one year. Windows NT-based Viewpoint client/server software displays a 3-D facility floor or a directory tree with defined zones for quick and easy location searches. The system uses standard open architecture and can be integrated with most databases running on Windows NT. Another example using RFID is Wal-Mart, which enables information about its products to move through the supply chain more quickly. Wal-Mart has indicated that it wants all of its suppliers to be shipping RFID-tagged pallets by 2006. Basically, RFID technology is at the heart of its operational systems to improve inventory management and forecasting. It is part of Wal-Mart's contribution to retail standards.

Pinpoint has implemented its system in the plant of a major U.S. auto manufacturer. Although an ERP system is used to plan production and work orders are sent daily to the production floor, it is very important at each stage of production line to make sure that all parts are physically there. In this very-high-velocity environment, with cars coming off the line at a rate of 50 per hour and with inventory coming in on a JIT basis, parts rarely make it to storage. Without the Pinpoint system, visibility is partial, even with a fully functioning warehouse management system. For example, parts may be somewhere in the facility on a forklift or they are often misplaced. All in all, the Pinpoint system allows continual visibility from the time the component is tagged at receiving and requires no human interaction to record or scan locations beyond that point (Navas, 2000).

Optimal KM/WM System Application in Manufacturing

Optimal KM/WM system applications in manufacturing are just becoming operational since there is a trend away from a build-to-stock basis to a build-to-order basis. The economics, as discussed earlier in the chapter, can be substantial for a typical company. Lexmark International Inc., for example, is using the Internet to manage and grow its operations for its worldwide operations. Essentially, this company utilizes the Internet to manage and grow its ink-cartridge manufacturing so that wise decisions are being made at all of its plants worldwide. It is capturing manufacturing data on Unix servers; and generating reports in HTML for viewing on thin clients. Lexmark is using Camstar's InSite manufacturing execution system (MES) as its common platform for capturing and processing data at all of its plants. Lexmark plans to reduce the number of defective products or unusable manufacturing leftovers, which many manufacturers call scrap. Plant engineers can now monitor ink cartridges throughout the production process and can also measure product quality, with real-time access to information on product tests via the Internet. An engineer, after doing some studying of a returned or defective product, could come in through a thin client and put the entire batch on hold anywhere in the world (Alexander, 2002). The capabilities of using an optimal KM/WM system-operating mode could be big cost savings to Lexmark in the coming years.

Summary

Because the manufacturing function has and continues to witness many exciting developments, this function can reap many benefits by employing optimal KM/WM systems. As noted initially in the chapter, management must rethink its manufacturing operations for optimal decisions. This rethinking is in line with the employment of advanced manufacturing methods, newer planning and execution systems, such as enterprise resource planning and manufacturing execution systems plus optimization techniques that center on mathematical and statistical models. For these newer advances to be effective within an optimal KM/WM system operating mode, there is a need for manufacturing management to take an enlarged view of wisdom by connecting their points where deemed necessary. Next, the essential elements underlying the development of an effective manufacturing model for optimal KM/WM systems were explored in some depth. In turn, a manufacturing model and its sub-models were set forth within such an environment. To obtain a better understanding of production planning and execution, its essential parts were treated in some depth whereby manufacturing operations were connected via points of wisdom for making decisions currently as well as over time. In this exposition, there was a tie-in with the other functional areas of a typical company. Lastly, an optimal KM/WM system application in manufacturing was presented.

References

Alexander, M. (2002, March 12). Factory floors go online. *Internet Week*, 21-22.

Bacheldor, B. (2003, March 3). Supply demand. *Information Week*, 47-58.

Baljko, J. (2003, March 3). Built for speed. *EBN*, 25-28.

Chabrow, E. (2005, June 27). Computers get the meaning. *Information Week*, 61.

Konicki, S. (2002, April 1). Reviving up. *Information Week*, 16-18.

Leibs, S. (2003, February). ERP software, full speed ahead. *CFO*, 63-66

Navas, D. (2000, June). Automotive on the cutting edge. *IDS*, 36-48.

Navas, D. (2000, October). Internet dreams: Manufacturer's trembling embrace. *IDS*, 32-44.

Neil, S. (2004, October). Going beyond multinational manufacturing. *Managing Automation*, 48-50.

Neil, S. (2005, July). MOM is your new MES. *Managing Automation*, 38-41.

Rosmarin, R. (2004, October). Your custom car is ready. *Business 2.0*, 150.

Schlosser, J. (2004, December 13). Cashing in on the world of me. *Fortune*, 244-250.

Schwartz, E. (2001, March 26). Nissan drives down costs with optimization software. *InfoWorld*, 62.

Stackpole, B. (2005, May). Smooth ride to PLM. *Managing Automation*, 48-50.

Chapter X

Real-World Optimal KM/WM System Applications

Issues

- To take a last look at creative thinking and problem finding that underlies "what needs to be done" in a typical optimal KM/WM system
- To tie in new business models with the ability of a company's decision makers to judge soundly over time
- To review computer software that is useful in optimal KM/WM system applications
- To set forth real-world optimal KM/WM system applications in the areas of corporate planning, marketing, finance, and manufacturing

Introduction

The focus of this chapter, including selected parts of Chapters VI through IX, is on the application of optimal KM/WM systems to various types of organizations in different fields. As discussed in Chapter I, some organizations are well along the way to this new type of operating mode since they have taken a *holistic*, that is, *comprehensive*, *approach* while others have taken a *segmented*, that is, *functional*, *approach* with the idea of finally taking a comprehensive approach. Also, some organizations have taken a *functional approach* to some area of the organization that can make or break it. Hopefully, they will eventually take a comprehensive optimal KM/WM system approach. The real thrust of undertaking the development and implementation of optimal KM/WM systems is to assist a company's decision makers in judging more soundly about their decisions about what needs to be done today and tomorrow.

In this application chapter of optimal KM/WM systems, there is initially a last look at creative thinking and problem finding. Many times, a creative approach may signal the need for new business models that are more reflective of the times. Next, a review of appropriate computer software that is useful in optimal KM/WM systems is set forth. The main focus of the chapter is on effective applications in the areas of corporate planning, marketing, finance, and manufacturing. Additionally, an application is given as a holistic approach at the end of the chapter. Overall, the concepts underlying optimal KM/WM systems are found in the applications set forth in this chapter.

A Last Look at Creative Thinking and Problem Finding that Underlie "What Needs to be Done"

The materials in this chapter, like previous ones, center on the basics of optimal KM/WM systems for what needs to be done. In the process of how the organization conducts its business and why it is done, creative thinking and problem finding can be extremely helpful to decision makers for optimal decision making. Organizational business activities include corporate planning, marketing, finance, manufacturing, and human resources as well as the many relationships and interactions that exist among them. Creative thinking and problem finding help company decision makers to cut through the haze and fog of the ever-changing business environment by employing sound judgment about operations. It allows decision makers to get the big picture versus some part of it. From this improved visibility standpoint, decision makers can better assess how to do more with less. In a similar manner, this improved visibility can be applied to a company's employees, that is, the success of an organization depends on how well individual employees connect with individual customers. A person-to-person business puts employees in the middle and views processes and procedures, information technology, and knowledge-sharing systems as resources that employees may use to be more effective when dealing with customers. By having a company's employees very connected to its customers, the end result is a heightened visibility of where the customer is headed. From this broad-based view of customers, a company can better anticipate their needs before they arc actually aware of them.

One last point should be made about creative thinking and problem finding that underlie optimal KM/WM systems for what needs to be done over time. Opportunities are limited by the imagination of the decision makers. Endless opportunities go hand in hand with allowing decision makers the ability to explore unusual and unexpected ways of solving problems and developing new opportunities for optimal decision making. For example, expectations can take the form of a CEO or an e-business executive who conjures up Web systems he or she thinks can be developed to exploit new business opportunities. In a similar manner, a marketing executive who focuses on newer approaches that result in lower prices for higher-quality products supported by better service is another example. The point being made here is that the inconceivable may be conceivable is viewed from a different perspective for improving the bottom line. Doing more with less from a heightened visibility standpoint within a new environment should not impair a company's financial results but improve them over time.

New Business Models Provide a Means for Adapting Company Operations to Changing Times

Creative thinking and problem finding are an integral part of developing new business models for any type organization. Depending upon the industry, new business models are needed more often in a high-tech industry where changes are the order of the day versus more traditional industries where changes come over a longer time period. Typical new business models software was given in Chapter IV. In many cases, however, companies find it easier to customize their business models to fit the specifics of their organizations to adapt to changing times.

From a very broad perspective, new business models found in optimal KM/WM systems center on integrating core data, information, and knowledge with business intelligence and optimization results to illuminate forthcoming problems and promising opportunities for implementation. These models can assist corporate decision makers in getting to know what their customer needs are before the customers are aware of them. As such, these models fulfill the strategic requirements of corporate decision makers at the highest level so that their decisions are based on sound judgment. Additionally, their use can assist corporate decision makers in getting a solid grasp on a company's overall direction. Feedback built into these new business models can help corporate decision makers to determine the need for newer products and services that are more in tune with the times. In the final analysis, corporate planning and its relationship to the organization's functional areas allow corporate decision makers the ability to connect "points of wisdom" for optimal decisions over time.

Computer Software that is Useful in Optimal KM/WM System Applications

Related to effective new business models is appropriate optimal KM/WM systems software to facilitate their usage. As discussed at length in Chapter IV, there are a number of software packages that are helpful in developing and implementing optimal KM/WM systems. These are:

- *optimization software* — useful in searching for the best solution for well-structured problems or near best solution for poorly structured problems.
- *goal programming software* — helpful in solving for a single objective with multiple subobjectives or multiple objectives with multiple subobjectives.
- *product lifecycle management software* — useful in gathering, relating, appraising, and projecting all information, knowledge, and intelligence related to a product or service over its life cycle.
- *predictive analytics software* — helpful in getting a handle on customer behavior as well as forecast product demand and related market dynamics.
- *knowledge discovery (data mining) software* — involves the extraction of hidden predictive information and resulting knowledge from large databases.
- *data visualization software* — has become the method of choice for understanding many complex business issues and activities.
- *virtual reality software* — is helpful to assist decision makers since it uses size, shape, and color in a 3-D landscape to convey useful knowledge that has been highlighted from extensive analysis.

In addition to the above, there are other software packages found in Chapters VI through IX. More specifically, they include business planning, balanced scorecard, behavior mapping, CRM analytics, alternative to budgeting, financial analytics, smart technology, supply chain management, enterprise resource planning, and manufacturing execution systems. Overall, there is a wide range of computer software that is very helpful in developing and implementing effective optimal KM/WM systems, some of which are explored below.

Effective Real-World Optimal KM/WM System Applications

A number of real-world KM/WM system applications were given previously in the text. They are:

- corporate planning application – Salesforce.com® (Chapter VI)
- marketing application – Wal-Mart® (Chapter VII)
- finance application – Charles Schwab (Chapter VIII)
- manufacturing application – Lexmark International Inc.™ (Chapter IX)

In this chapter, a number of other applications that can be related to these chapters are given below.

- corporate planning application – Intuit®
- marketing application – Procter & Gamble
- finance application – Deere & Company
- manufacturing application – Kellogg Company™

Additionally, there is a master case study, which is an:

- overview application – General Electric Company

It should be noted that Dell Computer™, including Procter & Gamble, were set forth in Chapter I.

Focus on Corporate Planning: Intuit®

A successful corporate planning application can be found at the software maker Intuit. Steve Bennett was installed as its new chief executive officer in January 2000. Upon his arrival, he found a company losing money, its technology was outdated for the times, and its execution was extremely slow. In addition, nothing was documented like it should have been. However, he inherited two important strengths: a fiercely loyal customer base and three of the most powerful brands in retail software. They included Quicken, the personal finance program that is all but synonymous with Intuit®, TurboTax, and QuickBooks® which is the accounting program for small businesses.

As a starting point, Bennett initiated a corporate plan that would get the company growing again by concentrating on the company's areas of strengths. When Bennett came in, the QuickBooks® group was a slow-growing unit with great potential. The standard version of QuickBooks was designed for companies with fewer than 20 employees, yet 5 to 10% of its most loyal users were larger — some had as many as 200 employees. In fact, more of them used QuickBooks® than brands of software intended for companies their size. Thus, Bennett realized that there was room for more than just an expanded version of QuickBooks.

In light of this important fact, Bennett hired Larrie Norrington from General Electric, who was given the mandate to make Intuit the SAP or Oracle® of small business. The company would offer software and services for a wide variety of organizations, from the smallest shops to those with a couple of hundred employees. Intuit would help not just with accounting, but also with payroll and benefits, keeping track of customers, and managing computer systems. It would also customize its software for specific kinds of businesses, like accountancies and construction companies. The initiative would be called "Right for My Business." Within a short period of time, QuickBooks® had sliced the accounting market 25 ways, with special editions for the smallest of the small companies and larger small companies, including specific versions for retailers, distributors, contractors, and nonprofit organizations.

From another perspective, Intuit opened QuickBooks®'s source code to independent software developers. The developers write highly specialized applications for specific businesses. With an open interface, they can easily tie their programs into QuickBooks® and other Intuit software, creating a kind of small business enterprise-resource-planning package. In view of the above approaches to open up opportunities for its customers and software developers, Intuit has been able to fashion a comprehensive approach to software that has not been beaten by any competitor at this point (Nee, 2003). Additionally, Intuit has been quietly acquiring companies over the past few years to expand into new markets and add bigger companies to its customer base (Kontzer, 2004). Overall, Intuit® has connected points of wisdom in corporate planning and related activities that serve all stakeholders needs.

Focus on Marketing: Procter & Gamble

In June 2000, Procter & Gamble elected A.G. Lafley as its chief executive officer. Diligently and methodically, he immersed himself in this global organization and spread his message that P&G has to focus on big brands, markets, and customers. In addition, he spread the message that P&G, to win with powerful discounters, must slash costs and reinvest savings in marketing and product design. Declaring that acquisitions are somewhat risky, Lafley gained fame as the best organic-growth person in the consumer products industry. Basically, Lafley has led a turnaround of P&G's fortunes.

As a way of establishing a means of bringing marketing activity together in a new and different way not found in the past, Procter & Gamble has utilized a PLM approach. As noted previously in the text, the PLM approach will eventually evolve into a venture analysis modeling approach. To P&G, PLM is not about investing in a packaged suite of applications or even deploying a new enterprise software category. But rather, PLM is a corporate philosophy designed to deliver a significant reduction in product development costs, foster reuse, and shorten the consumer goods manufacturer's time-to-market. It focuses on an overall architecture to make sure that all parts fit together. It also means getting corporate-wide sponsorship and buy in on a common marketing model that encompasses all product-related activities and establishing an integration strategy that can link up existing and new systems that span departments as well as reach out externally to suppliers and customers.

Essentially, Procter & Gamble identified five areas it wanted to address as part of its overall strategy. The first was already underway — PDM (product data management) — for managing all information associated with product development and making it available to the right people in the right version. The other four focus on common business processes across its different brands: (1) product portfolio management for determining what to bring to market at what time; (2) collaborative product design for managing an initiative end-to-end throughout all phases of its lifecycle; (3) customer-needs management for ensuring requirements are seen throughout the design and materials-sourcing stages; and (4) materials sourcing for handling the creation, routing, and approval process of specifying materials (Stackpole, 2003). Overall, Procter & Gamble has embraced PLM and is working diligently on its implementation. In turn, the company will be implementing a venture analysis modeling approach that will maximize the company's resources over time. Essentially, both will be complementary to an effective supply-chain optimization system.

Just recently, Procter & Gamble and Gillette have agreed to a merger. Both companies have identified significant cost synergies in this combination — more than two-thirds of the value creation. P&G has always delivered on cost synergies. When important growth synergies are added, like China, Brazil, and India, this is where P&G and Gillette together can achieve critical mass. There are not many acquisitions, large or small, that have all these things coming together (Sellers, 2005). This synergy between the two companies can be very beneficial to promoting optimal KM/WM systems, especially for a holistic approach so that the right hand knows what the left hand is doing.

Focus on Finance: Deere & Company

In the area of finance, the Deere & Company is a good example. When the economy and the stock market are at their peaks, customer financing seemed to be a smart move. Today, the balance sheets of companies-turned-banks are sagging under the combined weight of leverage and defaults. Many equipment makers are not too open about the extent to which they use financing to subsidize sales. But at Deere & Company, the company levies a capital charge on its equipment divisions for financing that the parent provides, something that most other equipment makers do not do. Many companies have gone astray where they have just assumed that they will be paid back their money and that financing will add to revenue. Deere charges its equipment divisions the full carrying costs of financing dealer receivables, including a market return on the equity, and those charges come out of their operating profit. That is different because other companies charge right out of operating profit for those carrying costs.

Deere's financial conservatism does not end with the transparency of its financing. The company has marketable securities and cash on hand of almost $3 billion and that it refrains from financial engineering which raises concerns that offset short-term benefits. For instance, Deere limits its use of receivables securitization to more than 20 to 30% of total debt, whereas some companies are securitizing as many of their receivables as they can. While such deals accelerate income, Deere rather earns income in a more steady fashion. Such wise thinking has put Deere in a good position. A credit downgrade by the rating agencies recently, taken out of concern over unfunded pension and other postretirement liabilities and the economic downturn's potential toll on farm-equipment sales, all but cost the company its ability to issue inexpensive commercial paper. Yet Deere's interest expense has remained virtually unchanged. In effect, there has been no significant drop in Deere's results stemming from its rating downgrades. In an environment so averse to debt, this wise approach to financing assists Deere & Company not only today, but also will assist it in the future. Overall, a conservative approach to reporting financial results stands very well with investors as well as makes internal reporting easier to prepare on a monthly basis and on a daily basis (Fink, 2003).

In addition, Deere & Company has been successful in redoing its level of inventory by a substantial amount. Before turning to SmartOps (a supply chain optimization software vendor), Deere used a "crude rule of thumb" to plan target inventory levels for each dealer. The goal was to maintain 30% of annual sales in inventory for each dealer. However, this rule of thumb did not take into account variability, that is, seasonality or the specific requirements of individual dealers. Also, replenishment cycles represent another important variable. If a product takes two weeks to replenish, the variability might be so great a dealer could run out. If the same product took two days to replenish, the dealer could carry much less inventory because there is less variability in two days than in two weeks. Thus the question becomes: "How much inventory is needed to cover the variability in both supply and demand?" Essentially, SmartOps loads the data from three John Deere plants and 25 dealers into its MIPO (Multistage Inventory Planning and Optimization) software. By designing a model that forecasts what products should be at which locations each week, the MIPO software showed how Deere could improve service levels while maintaining less inventory. Since the project began in 2002, Deere has reduced inventory to the tune of $1 billion (Schwartz, 2005).

Focus on Manufacturing: Kellogg Company

Carlos Gutierrez of the Kellogg Company, since becoming its CEO several years ago, has transformed this volume-oriented company into an innovative, profit-focused organization. He has breathed new life into brands, like Frosted Flakes and Special K, regaining the top spot in the U.S. cereal market. Also, he acquired cookies-and snack-maker Keebler to expand the company's product line and distribution capability. The net result is that Gutierrez has put some snap, crackle, and pop into Kellogg's stocks, out performing its competitors for the last several years.

In terms of the company's new business model, it centers on Volume to Value, that is, sales are grown by shifting to higher-margin products. These include

Special K, which appeals to weight-conscious women; Kashi, which targets the health food set; and Nutri-Grain bars, which lets the customer have breakfast on the go. In contrast, cereals such as Kellogg's corn flakes have lower margin because they have been so widely copied that they are essentially commodities. Furthermore, the new business model has allowed the company to alter its daily tracking system to record dollar sales not pounds, and to overhaul its bonus plan to reward profits and cash flow not volume. Gutierrez called for cutting excess capacity, namely, shutting down the 93-year-old South Plant in Battle Creek, a mile from Kellogg's headquarters.

Additionally, Gutierrez stressed improving relations with Wal-Mart. In the past, Wal-Mart did not take kindly to the company's inflexible policies or pricing and promotion. He shifted top talent to the Wal-Mart account team, which put more of Kellogg's products on Wal-Mart's shelves, and tapped long-standing Kellogg market knowledge overseas that had helped the company expand into new regions. Going one step further, Gutierrez hired more than 200 direct-sales representatives for the U.S. cereal business, replacing inefficient middlemen and accelerating distribution of new products in supermarkets like Kroger. Overall, improved methods have provided the means for Gutierrez to increase the company's profitability and its productivity today and tomorrow. Essentially, he has developed and implemented some of the basic elements found in optimal KM/WM systems that are related to its new business model. Gutierrez has been provided the means to connect "points of wisdom" for optimal decisions that were needed to bring the Kellogg Company up to current times (Boyle, 2004).

Focus on Overview: General Electric Company

A holistic approach to optimal KM/WM systems is currently underway at the General Electric Company. Three years into his chief executive officer role, Jeffrey Immelt is undertaking a radical reinvention of one of the flagship operations of American industry. Immelt is de-emphasizing some of the tenets that his predecessor, Jack Welch, made famous. Instead, Immelt has staked GE's future growth on the force that guided the company at birth and for much of its history. He is after new, breathtaking technological innovation. Essentially, he has concluded that to move GE forward, his managers must view the company not so much as a collection of huge multibillion-dollar businesses, but as a network of entrepreneurial, Silicon Valley style tech start-ups.

His mandate to his managers is that General Electric needs to lead on the far technological frontiers of markets from clean energy to medical diagnostics to nanotech to security to jet-propulsion systems. And he wants to harness that newer innovative mentality to a more developed and systematic marketing effort that GE has not done before. The bottom line is that Immelt wants GE to be an entirely different company which is more entrepreneurial and more science-based as well as capable of generating much more growth from its own internal operations than by simply acquiring other companies as in the past.

From an overview standpoint, growth targets set by Immelt require that General Electric generate more than $9 billion in new revenue annually from internal operations alone. That is like adding the combined businesses of eBay, JetBlue, MGM, and Starbucks. Few companies in the past have been able to consistently generate that kind of sales revenue without acquisitions. The challenges of getting there by igniting internal innovation are extremely large but, in many ways, are no different from the issues all companies face in trying to unleash latent entrepreneurial power. How Immelt is going about it contains lessons for managers everywhere (Schonfeld, 2004). Needless to say, he and his high-level managers need far reaching wisdom for resolving the company's forthcoming problems and taking advantage of the many promising opportunities. Essentially, this means seeing through important relationships so as to connect "points of wisdom" about what needs to be done over time. Such an approach requires the assistance of advanced computing power, that is, optimal KM/WM systems, to achieve such far-reaching goals. Such systems will allow his decision makers to judge the soundness of their decisions about what needs to be done for growing this large learning organization.

More recently, General Electric has reorganized its 11 businesses into six industry-focused groups. In addition, GE named three new vice chairmen and reaffirmed double-digit growth for the coming years. These moves signal a changing of the guard and stepped up efforts by one of the world's largest companies by market value to tap into fast-growing markets in developing countries. The changes also are designed to retain top executives who have been sought after by other major companies and will save GE many millions. These moves give the company a chance to gain market share and to do so in a cost effective way. Overall, this organization provides the company with the ability to realize more of its potential, especially with the assistance of optimal KM/WM systems today and tomorrow.

Summary

Initially, the chapter took a last look at creative thinking and problem finding since they are an integral part of optimal KM/WM systems. A starting point for these systems was covered under new business models. In turn, an overview of computer software that is useful in such systems was presented. Next, the focus was on presenting real-world applications of optimal KM/WM systems that demonstrate how organizations are adapting to changing times. This new enlightened approach to information technology is a willingness to start not with the question "What do you think we should do?" but with the question "What do we need to do?" From this very broad perspective, optimal KM/WM systems give a company's decision makers the ability to judge soundly so that optimal decisions can be made about what needs to be done.

References

Boyle, M. (2004, September 6). The man who fixed Kellogg. *Fortune*, 218-226.

Fink, R. (2003, March). What goes around. *CFO*, 44-52.

Kontzer, T. (2004, November 22). New horizons. *Information Week*, 28-34.

Nee, E. (2003, June). The hottest CEO in tech. *Business 2.0*, 86-92.

Schonfeld, E. (2004, July). GE sees the light. *Business 2.0*, 80-86.

Schwartz, E. (2005, April 25). Supply-chain victory. *InfoWorld*, 12.

Sellers, P. (2005, February 21). It was a no-brainer. *Fortune*, 97-102.

Stackpole, B. (2003, March). Demystifying PLM. *Managing Automation*, 20-21.

Section V:

The Impact of the Future on Optimal Knowledge Management/Wisdom Management Systems

Chapter XI

Future Developments that Impact Optimal KM/WM Systems

Issues

- To explore a fourth-dimensional view as a requirement for a true optimal KM/WM system
- To look at the future developments of computer storage, networking, and software as they affect a well-designed optimal KM/WM system
- To examine future developments in corporate planning, marketing, finance, and manufacturing as they affect optimal KM/WM systems
- To explore the need for continual support of optimal KM/WM systems for what needs to be done over time

Introduction

Today, companies are being pressed to respond to customer needs and competitive threats in days and weeks instead of months or years. Products and projects that could linger for six to twelve months just a few years ago now need to get out the door in a much shorter time frame. And it is not just multinationals or global corporations that are being faced with shortening time frames. Almost any company, from a small company up to the world's largest corporation, is at risk of being "Amazoned" by a more nimble, e-business-enabled competitor. The success of companies will be measured by how well they have leveraged e-business applications to differentiate themselves from the competition. How well the company can respond to changing times is paramount. The implementation of optimal KM/WM systems can make a significant difference in how companies respond to changing times.

Initially, the chapter examines a fourth-dimensional view that underlies future developments that really impact optimal KM/WM systems. Future computer storage, networking, and software developments as they affect optimal KM/WM systems are discussed, followed by future considerations for developing and implementing a well-designed system. Future developments in the areas of corporate planning, marketing, finance, and manufacturing are examined and their tie-in with improving a company's decision makers wisdom. Finally, there is a discussion on the continuing need for supporting optimal KM/WM systems to assist a company's decision makers over time.

A Fourth-Dimensional View is a Requirement of True Optimal KM/WM Systems

Currently, many information systems are essentially real-time systems that monitor specific operations for obtaining good solutions to current operations. Typically, these systems focus on a *three-dimensional* viewpoint. What is needed is a *fourth-dimensional* view that is over time. Effective optimal KM/WM systems enlist all of the pertinent inputs and expert assistance down some path to a wise and productive solution to a problem or an opportunity for the decision maker. The continuing turbulence of the times, but more importantly in the future, dictate that decision makers undergo a continuing "mind shift" so as to judge organizational operations soundly over time. As was seen in this text, the real essence of optimal KM/WM systems center on decision makers taking a holistic, that is, "big picture" view of the organization's operations by connecting "points of wisdom" about what needs to be done to grow the learning environment over time.

Future Computer Storage and Networking Developments that Affect Optimal KM/WM Systems

As in the past, an effective way to develop and grow an optimal KM/WM system today and tomorrow is to create a computer storage infra-structure within an appropriate computer networking structure that is supported by newer computer software. By bringing together data sources via computer networking from all parts of an organization internally and externally that is utilizing the latest software, decision makers are able to get a handle on the interworkings of an organization's operations and what needs to be done to keep it current. For many situations, this requires thinking beyond the company's products and services by getting involved with their customers' present and future operations and helping them redo their work and their way of conducting their business. Essentially, a future operating environment must reinforce the present computer storage and networking infrastructures already in place, tying them together with the appropriate software and delivering that which is greater than the sum of all parts for itself, its customers, and its trading partners. In turn, this future operating environment must facilitate the operations of all parties in a way that is not possible if each party operated alone.

In addition, there should be provision for supportive collaborative computing that supports current and future workflow technology. Decision makers should be able to obtain information, knowledge, intelligence and optimization results from anywhere. They should utilize an easy-to-use approach without regard for their sources and types to optimize a company's operations so that wise decisions can be made. Basically, decision makers need to think in terms of how their companies tie-in with global concerns today and tomorrow. Otherwise, competitors will figure out what is best and leave these companies in dire straits.

Future Software Developments as They Affect Optimal KM/WM Systems

Future software developments for optimal KM/WM systems will focus on all or, at least, most of an organization's business processes such that a company's points of wisdom can be tied together in an optimum manner. The ability to bring complex business processes together from beginning to end is a formidable challenge facing system developers today and in the future. Because typical future software developments are directed to aid in optimizing a company's business processes, their focus is aimed toward helping decision makers in judging more soundly the complexity of an organization's operations. In order to accomplish this goal, forward-looking enterprises of the future will pull real-time information, knowledge, intelligence, and optimization results from points of wisdom sensors in the process, that is, from manufacturing, supply chains, even to retail stores and customers. This flow that comes from an armada of organizational systems will feed compressed product and service life cycles driven faster and faster by the business need to increase profit margins. Collaboration among applications and systems along with the needs of decision makers will become mission critical. The high tech hardware and software companies are expected to respond with an explosion of newer techniques, software tools, and new business models. The bottom line is that future software developments of all kinds for optimal KM/WM systems are needed to assist decision makers in their pursuit of judging soundly a company's operations at all times today and tomorrow.

Future Considerations for Developing and Implementing a Well-Designed Optimal KM/WM System

The design and implementation of a well-designed optimal KM/WM system, as noted in Chapter V, requires taking a fresh and new look at the whole development process. In an ever-changing, dynamic environment of today and tomorrow where things are far from static, projects can become outdated before they can be delivered. As the business environment has moved more toward e-commerce and the Internet plus more optimal KM/ WM system applications, there is need for a newer direction to an organization's operations. Today and in the future, the focus is on designing and implementing optimal KM/WM systems that include flexibility in their basic design.

Within a flexible environment, systems designers must address the question of buy versus build. This evaluation pits buying (typically a packaged, third-party application) against building (typically hand coding) with the winning path usually representing the lesser of two evils. Generally, neither the buy nor the build response can truly meet the complete needs of decision makers since both of these approaches can result in inflexible, static applications that can quickly become obsolete in the face of change. By utilizing either one of these approaches alone, businesses run the risk of facing significant short- and long-term challenges that are not easily overcome. Hence, a combination of buy and some custom alterations to new optimal KM/WM systems is generally necessary to meet the changing needs of decision makers.

Future Corporate Planning Developments Affecting Optimal KM/WM Systems

Going beyond the contents set forth in the chapter for corporate planning, future developments are coming into play daily and periodically to assist corporate planners. They include the digital dashboard and virtual worlds, both of which are explored below. Essentially, a *digital dashboard* is a continuously updated online display of a company's financial ratios and statistics. For example, at General Electric, "digital cockpits" give 300 managers access to company essential information — on desktop PCs and PDAs. The previous system required dozens of analysts to compile information and knowledge and send it up the line. In contrast, digital dashboards have reduced their ranks to six. To get the project launched, the project director asked several senior managers — from quality assurance, manufacturing, and information systems — to decide what numbers each division would contribute. The project director then hired the computer professionals to enforce data input. The numbers generated would be a day old, but that was a fair compromise. Live data would have cost ten times more and the payback was not there. It should be noted that the biggest challenge of all was about "changing the culture" so everyone has a common way to look at the business (Tedeschi, 2003).

From another view, *virtual worlds* allow decision makers to explore how much money is spent and where along with useful financial ratios and statistics. Virtual reality (VR) software is useful in creating a walk-through scenario of how a company spends its money, that is, on capital projects found within a PLM operating mode. Particular attention is given to flagging low-priority projects that were overfunded. Capital project activities can be assigned short or tall bars by their spending levels, and color-coded from red to green, to show priority. Exploring a 3-D landscape, decision makers are able to determine which activities to view, touch, and manipulate for further analysis. In effect, decision makers can step back to see how resources are consumed for each area or project of a company as well as for the entire company and all projects (Thierauf, 1995).

Future Marketing Developments Affecting Optimal KM/WM Systems

Future marketing developments affecting optimal KM/WM systems include: (1) further improvements in partnerships, (2) increasing sales volume, (3) importance of customer retention, and (4) focusing on indicators of quality over the long term. In reference to the first item, there will be a continual recognition that partnering companies are not willing to throw everything away and start anew. Instead, they continuously provide a host of enrichment for the values desired by all partners. In terms of the second item, steady sales growth can be spurred not by dropping price, but by the collective efforts of increased competition. That is, the massive force of competitors in alliances or cross-licensing agreements could help move the sales of products and services to a higher level. Third, the importance of customer retention needs to be examined in terms of an expanded view of wisdom. That is, companies need to think in terms of customers who not only use and talk about products and services, but also how they serve as role models for potential purchasers. Taken together, these activities can significantly enhance the value of customers to a company.

Although many developments will affect optimal KM/WM systems in marketing, a most important one that underlies the above and others is the fourth item — involves the perceived quality of the product or service, no matter how subtle it may be. Basically, companies need to determine what story they want to tell their customers, and then ensure that their employees, including trading partners, consistently show customers evidence of that story. Eventually, this is the story that will find its way to chat rooms on the World Wide Web and one-to-one conversations among current and potential customers. In just about any organization, the clues emitted by people and merchandising circumstances tell a story to customers. An effective approach of what needs to be done organizationally is designed to tell the intended story for today and into the future (Berry & Bendapudi, 2003).

Future Finance Developments Affecting Optimal KM/WM Systems

Future finance developments affecting KM/WM systems, as in the past, will focus on profitability rather than wait for revenue growth to solve the profitability problem. An integral part of this underlying approach to maximizing profitability includes the following: (1) a further enhancement of the value proposition concept (stressed throughout the text), (2) a forward-looking focus on the big picture so the company is always pursuing the appropriate financial objectives and goals for the times, (3) a further move away from budgets as stressed in Chapter VIII, and (4) a continuing utilization of financial analytics. Each of these areas is discussed in the material to follow.

In terms of the first item — a further enhancement of the value proposition concept — there will tend to be a future emphasis on what value society places on a company's goods and services versus just the consumer. Although current information systems are designed to hold the individual manager accountable for his or her actions, the future focus will be more on coordination and knowledge sharing within the company as well as outside the company. The net result will be a win-win situation for society and the company. A socially responsible firm has a better chance of getting the support of the community when times are bad. Hence, a company, by being a good citizen, has a better chance of survival in the long run.

Related to the value proposition concept is the Value Measurement and Reporting Collaborative (VMRC), which believes value, is defined not only in monetary units, but also in objects, ideas, events, or processes. The VMRC, in which the American Institute of Certified Public Accountants (AICPA) participates, is a global effort of the accounting profession to help board of directors, senior management, investors, and other stakeholders make better strategic decisions using value measurement and reporting. Its member's say a company's worth exists not only in its present operational value as found in current financial statements, but also in its potential to create future value. The challenge of VMRC is taking up the development of principles and criteria that characterize value measurement and reporting worldwide (Drozal, 2004).

The second item — a forward-looking focus on the big picture so the company is always pursuing the appropriate objectives and goals for the times – ties-in with the transformation of the corporate finance function, driven in part by newer technological advances. Essentially, these newer technologies have freed financial decision makers from routine information gathering tasks and have allowed them to focus on the big picture of its own operations as well as that occurring outside the organization that are essential to its financial health. From this broad perspective, financial decision makers must also be good at assessing nonfinancial measures that impact the organization. The big picture for financial decision makers in the future center on merging both financial and nonfinancial measures of corporate performance within a problem finding operating mode. A good balance between these two measures is a must for future financial decision makers to be effective over the long term.

For the third item — a further move away from budgets as noted in a previous chapter — there will be an emphasis on using benchmarks of successful companies in their respective industries and the competition of peer groups throughout the organization so that the best can be accomplished with a minimum of costs for the highest quality. By the same token, other nonfinancial ratios, like better assessing the customer after the individual has acquired the product or service or getting the item to the market faster in order to be known as the market leader, will be stressed. On the other hand, future corporate financial measurements will focus on anticipated income-cost-profit ratios that should be experienced over the lifecycles of products and services. The bottom line is that by creating a lifetime value analysis for a company's products and services, the company's actual cash flow can be evaluated from a more realistic viewpoint.

For the fourth and last item — a continuing utilization of financial analytics — a robust analytic platform should make information, knowledge, intelligence, and optimization results available to all users, not just top-level decision makers. Basically, financial analytics should transcend everyone in the organization. For example, a senior-level executive needs intelligence on the company's latest sales figures to deliver a gross-aggregated view of its revenue goals for promoting prepaid services. Additionally, staff members benefit from analytics since it allows them the capability to review and improve their personal performance. The wisdom produced as a result of financial analytics should provide the underlying performance, scalability, and flexibility to grow and evolve as the organization's needs increase and change ever time. Overall, the above four future finance developments will go a long way to enlarging the scope of optimal KM/WM systems over time.

Future Manufacturing Developments Affecting Optimal KM/WM Systems

Although newer manufacturing techniques and software are found today, future manufacturing developments affecting optimal KM/WM systems are expected in the following areas. They include: (1) customer-centric manufacturing, (2) collaborative product commerce, and (3) a greater tie-in with a venture analysis modeling approach. In terms of the first item, because the Internet is the instrument of the many changes taking place in the manufacturing environment, a newer approach is about everything that happens after the order is placed. The focus is on order creation as customer specifications become more important and product configuration, flexible pricing models, and remote order management take over. In reality, the customer is taking over as chief architect of a manufacturer's future product development. From this view, there is a customer-centric approach to manufacturing operations not found in the past.

Related to customer-centric manufacturing is collaborate product commerce (CPC) — the second item of future manufacturing developments. Although manufacturers have developed computerized systems to help them streamline their business processes, increasingly, that meant sharing information, knowledge, intelligence, and optimization results with partners and suppliers in an e-commerce offering that optimizes order fulfillment, inventory management, and distribution processes, that is, supply chain management. To enhance their competitive advantage, some manufacturers are taking business-process integration initiatives a step further by bringing partners and suppliers in at the initial phase, that is, product-development phase. From this perspective, collaborative product commerce addresses the problem of managing and sharing the frequently changing facts and figures surrounding the design, development, and manufacturing of a product, thereby delivering significant benefits in the process to all parties.

Regarding the third and last item, a tie-in of customers and business partners and suppliers with a manufacturer can be extended further by a greater integration with a venture analysis modeling approach noted throughout the text. More specifically, a comprehensive venture analysis modeling approach looks at the product over its complete life. An important caveat here is that decision makers must know how "fresh" their input is, which means they have to know how current the information, knowledge, intelligence, and optimization results is upon which they are acting. Getting better control of these items that are stored out there on servers and using Web technology as the delivery mechanism is essential. In order to out-deliver and out-invent the competition, companies have to create facts and figures about customer preferences and turn that into rapid development/configuration. The bottom line is that a venture analysis modeling approach needs to move toward operational optimization for effective everyday wise decisions that tie-in with customers, partners, and suppliers.

Continual Support of Optimal KM/WM Systems for What Needs to be Done Over Time

No matter what approach is used in developing and implementing optimal KM/WM systems, there is need for some type of training of decision makers and their staffs so that they have an increased ability to judge their decisions more soundly over time. Typically, they need to know how to use the level of sophistication displayed on their screens. If hand holding of the company's personnel is undertaken, optimal KM/WM systems stand a much higher chance of being accepted because decision makers tend not to bother with a user's manual or even a user's pamphlet.

For effective continuing support for what needs to be done over time, essentially the project team becomes the support team. However, it usually takes fewer people to support the optimal KM/WM system than to develop it initially. The support team members should have a continuous interest in the system's success. Needless to say, the implementation of such a system can cause major changes throughout the organization in terms of altering the flow of data, information, knowledge, intelligence, and optimization results to and from the top of the company. Within this context, decision makers can spend considerable time undertaking the appropriate analyses necessary for optimizing the company's strategic-business direction. As the world-renowned management consultant Peter Drucker has said in the past, that which is measured ultimately improves. Updating this concept, an optimal KM/WM approach must provide a basis for learning and compounding of learning in order to create newfound wisdom for understanding and, in turn, optimizing a company's performance over time. The bottom line is that this approach becomes an important intellectual company asset that needs to be managed effectively over time. Wisdom that underlies a company's intellectual assets can be used to change the rules of competition so as to redefine the entire market. Unlike its competition, a firm can choose to rewrite the rules to its own advantage; thereby leaving would be competitors struggling to keep up. By supporting these newer systems today and tomorrow, a company can use them to maximize its intellectual assets. Proper continual support of optimal KM/WM systems, then, can be used to deliver significant, measurable improvements to an organization's "return on intellectual capital".

Summary

Future developments for optimal KM/WN systems have the capability of changing how the typical company views its total operations. As pointed out in Chapter I, *truth management systems*, which are not fully defined at this time, are expected to supplement optimal KM/WM systems sometime in the distant future. Truth, being in conformance to fact or reality, comes from understanding fully the way that connected points of wisdom come together logically. As such, the more widespread that optimal KM/WM systems become, there will be a move to develop the essentials of truth management systems. At that point in time, the capabilities of advanced computer technology will allow the implementation of such systems where computers are essential to one's well being on a day-to-day basis. In conclusion, although optimal KM/WM systems are expected to experience phenomenal growth in the years to come and represent the "pinnacle" of information systems development, they represent one large leap forward before the development and implementation of truth management systems.

References

Berry, L. L., & Bendapudi, N. (2003, February). Clueing in customers. *Harvard Business Review*, 100-106.

Drozal, F. A. (2004, December). The components of value measurement. *Journal of Accountancy*, 79-82.

Tedeschi, R. (2003, March). End of the paper chase. *Business 2.0*, 64.

Thierauf, R. J. (1995). *Virtual reality systems for business*. Westport, CT: Quorum Books.

About the Authors

Robert J. Thierauf (PhD, CPA, Professor *Emeritus*, Xavier University, Cincinnati, Ohio) teaches at Xavier University and has served as chair of the Management and Information Systems Department for 13 years. Formerly, he served as a staff auditor and consultant with PriceWaterhouseCoopers for six years. While at Xavier University, Dr. Thierauf has extensively published in the areas of information systems, management, and operations systems, with 33 books published to date and with many translated for foreign publication. Currently, he is researching and writing about new business models with an emphasis on a questionnaire approach. This approach will focus on assisting the organization's decision makers to become more flexible in reacting to the changing business needs.

James J. Hoctor (MBA, CPA, Adjunct Professor, Xavier University, Cincinnati, Ohio) is a manager of information systems for The Kroger Company Corporate Headquarters in Cincinnati. Having worked through various levels within the organization, he has gained extensive experience in the finance and information systems areas. Currently, he is working to develop newer approaches to procurement, having already helped to develop a real-time distribution solution. He also co-authored Dr. Thierauf's most recent publication and is currently researching and writing about new business models with an emphasis on a questionnaire approach. This approach will focus on assisting the organization's decision makers to become more flexible in reacting to the changing business needs.

Index

Symbols

A

B

W